Occupational Safety and Health for the Young Professional

Occupational Safety and Health for the Young Professional

Elliot Laratonda

Bernan Press

Lanham • Boulder • New York • London

Published by Bernan Press
An imprint of The Rowman & Littlefield Publishing Group, Inc.
4501 Forbes Boulevard, Suite 200, Lanham, Maryland 20706
www.rowman.com

86-90 Paul Street, London EC2A 4NE

British Library Cataloguing in Publication Information Available

Library of Congress Cataloging-in-Publication Data

Names: Laratonda, Elliot, author.
Title: Occupational safety and health for the young professional / Elliot Laratonda.
Description: Lanham : Bernan Press, [2022] | Includes index.
Identifiers: LCCN 2022018626 (print) | LCCN 2022018627 (ebook) |
    ISBN 9781636710549 (paperback) | ISBN 9781636710556 (epub)
Subjects: LCSH: Industrial safety—Vocational guidance.
Classification: LCC T55 .L36 2022 (print) | LCC T55 (ebook) | DDC 658.3/82—dc23/
    eng/20220711
LC record available at https://lccn.loc.gov/2022018626
LC ebook record available at https://lccn.loc.gov/2022018627

Contents

Preface

This is a handbook for aspiring, young safety professionals. This book is intended to be used as a textbook to supplement classroom instruction. The target audience of the book is late high school or early college students interested in the field of occupational safety and health. This is a perfect textbook for a class similar to Introduction to Occupational Safety and Health. The first three chapters of this book provide introductory material to orient the reader to the field of occupational safety and health, present fundamental concepts of hazard control and incident prevention, and introduce the Occupational Safety and Health Administration (OSHA) and its regulations.

The primary purpose of the book is to summarize and present OSHA regulations and standards in a way that is more easily understood and digested as compared to reading the standards in the code of federal regulations. The information presented does not require previous industry experience. Each chapter represents a primary occupational safety and health topic and related OSHA standards. The OSHA standards and information included in the book are a collection of topics the author feels are the most primary, fundamental, and core topics young professionals should understand who are working in general industry. This book is not a comprehensive outline of OSHA regulation. Various less common topics and standards as well as specific information within standards are left out of the book to present information in a way that is applicable to most readers and make the information more manageable to understand and comprehend. This is not a textbook about occupational safety and health program management or philosophy. It focuses on OSHA standard compliance. This is not a handbook for safety and health compliance in the construction industry or for employers working outside of general industry. This book is not a comprehensive resource related to health hazards or for industrial hygiene professionals. Lastly, this book does not cover topics relative to environmental protection and compliance. This is a textbook to motivate young professionals to practice within the field of occupational safety and health and allow for the understanding of common OSHA standards applicable to general industry.

In each chapter, in addition to the summary of the OSHA standard, there is a fictional narrative preceding the standard summary. Each narrative is about a workplace fatality, incident, or near tragedy related to the topic of the chapter. These narratives are inspired by real events and are intended to motivate interest in safety and health, so readers are inspired to prevent similar incidents, protect employees and save lives. Each narrative is intended to invoke sympathy and motivate readers to engage in the profession of occupational safety and health with genuine concern and sympathy for other people.

Acknowledgments

To my wife, Paige, thank you for your love and support during the many hours of writing this book, and thank you so much for contributing two chapters! I am very thankful to have a life partner who is so intelligent, kind, caring, and loving. I will always love you.

To my parents, thank you for your kind words, support, and interest during the writing process. I sincerely appreciate your interest in the endeavor and inspiration to achieve this goal.

Acknowledgments

Chapter 1

The Occupational Safety and Health Story and Fundamental Concepts

INTRODUCTION AND SCOPE

This chapter provides definitions of key terms commonly used in the field of occupational safety and health (S&H), a historical overview of the S&H movement, and fundamental concepts related to workplace S&H. The definitions provided are the opinions of the author. They are paraphrased from multiple sources and from personal experiences. The historical overview of occupational S&H is a summary of historical milestones as presented by the author. Accounts of these events are detailed in multiple sources. The fundamental concepts presented are the ideas compiled by the author after a decade of occupational safety and health education and industry experience and are considered common knowledge in the field of occupational S&H. For these stated reasons, references are not provided for most of the information.

The occupational S&H movement became significant as a result of the Industrial Revolution. The Industrial Revolution brought more hazards to the working environment, and society demanded a response from employers and the government. This led to the formation of workers' compensation programs in individual states as well as the creation of the Occupational Safety and Health Administration (OSHA), a government agency devoted to the protection of workers' S&H. Since the creation of OSHA, the field of occupational S&H has grown.

In the field of occupational S&H, the primary goal is to prevent workplace injuries, illnesses, and fatalities. To achieve this goal, employers implement S&H programs designed to identify, evaluate, and control hazards. This chapter provides (i.e., written plans) a fundamental overview of different approaches to identify hazards, describes how to assess risk that hazards pose, and covers different hazard control methods to mitigate hazards and reduce

1

risk. These fundamental concepts are crucial topics to understand as a young S&H professional.

DEFINITIONS

- Accident—an event that results in some sort of loss (e.g., injury, illness); by name, it is implied the event is unpredictable, unpreventable, and of short duration (which is not always the case in the field of S&H!)
- Hazard—anything that can cause harm or poses a threat to S&H, most commonly unsafe acts and unsafe conditions
- Incident—an event that may or may not result in a loss (e.g., an injury or illness including near-misses); a near-miss means the event "just-missed" and nearly resulted in a loss
- Occupational S&H—the professional field and practice of identifying and controlling hazards in the workplace to prevent incidents (e.g., injuries, illnesses, near-misses)
- Risk—a product of a probability rating and a severity rating for a hazard; the means by which to evaluate hazards; hazards present risk
- Safe—relatively free from harm
- Safety—the state of being relatively free from harm

HISTORICAL OVERVIEW

The occupational S&H storyline can be described in many ways. In this chapter, the narrative is summarized starting from the early 1800s spanning to the late twentieth century. Many people would begin the S&H story at the start of the Industrial Revolution, when work became more mechanical and automated rather than by hand. But, of course, people got hurt and sick at work before this. The foundations of each chapter in this book are based on OSHA regulations and standards. For that reason, perhaps the most significant event in this historical overview is the creation of OSHA. Despite government intervention for hazard control in the workplace, thousands of people are still fatally injured every day at work. For this reason, the field of occupational S&H will always be a very important profession.

    The occupational S&H movement traces its roots back to the global Industrial Revolution in the 1800s. England was the first to make notable progress toward safety in the workplace. In the early 1800s, child labor in England was common. Hours were long, the work was hard, and often the workplace was unhealthy and unsafe. After outbreaks of fever and illness in children in Manchester, England, there was public demand for better factory working conditions. In 1802, England passed the Health and Morals of Apprentices Act. This is commonly known as the beginning of governmental involvement in occupational S&H.

In the United States, during the 1800s, the Industrial Revolution led to more machines and more hazards in workplaces. This led to unsafe working conditions. In the first half of the 1800s, state-specific initiatives were passed including factory inspections, mine safety procedures, and requirements for machine safeguards. Then, in 1908, the United States passed Employer Liability Law. This made it possible, although extremely difficult, for railroad employees or families of employees to sue employers for negligence leading to incidents, injuries, or death. To win this legal battle, employees had to prove the incident was purely the fault of the employer (the employer was negligent). This was very difficult, because if the company could show the employee themselves contributed to the incident, or another employee contributed to the incident, or the employee should have assumed the risk involved in their job, then the company was off the hook.

The growing public outcry over hazards in the workplace and frustration over Employer Liability Law led to states pursuing their own systems for providing employees and their families compensation after being the victim of a workplace incident. In 1911, Wisconsin passed the first effective workers' compensation law. Today, all fifty states have some form of workers' compensation to provide injured employees or victims a portion of their wage and other benefits.

After the passing of workers' compensation law, into the late 1900s, work injury rates were on an upward swing. The US government wanted to intervene in order to protect their valuable resource of human workers, to preserve human lives, and to keep the United States economically viable in global commerce. In 1970, President Nixon signed into law the Occupational Safety and Health Act (OSH Act) of 1970. The Occupational Safety and Health Administration (OSHA) was created in 1971 to support the act, educate employers, set S&H standards, and enforce workplace S&H regulations. Since the creation of OSHA, most companies have integrated S&H as a core value.

Even after the creation of OSHA, in the 1980s, the world saw several major catastrophes that further motivated the importance of S&H rules and regulations in the workplace. In 1984, in Bhopal, India, a catastrophic failure led to a major release of a toxic gas called methyl isocyanate. Over 500,000 people were exposed to the deadly substance and about 2,500 people died. In 1986, the world watched in horror as the Challenger space shuttle burst into flames during takeoff. The same year, in 1986, the Chernobyl nuclear power plant, in Pripyat, Ukraine, exploded leading to the death of approximately 10,000 people. These are just a few examples of catastrophes that reinforced the importance of S&H in various industries.

The Industrial Revolution, public outcry over hazardous working conditions, OSHA, and a history of industrial incidents were major contributing causes to the field of occupational S&H as we know it today. The main purpose of S&H in the workplace is to prevent incidents and preserve the health and safety of workers. Employers and S&H professionals aim to prevent incidents by recognizing and controlling workplace hazards. The following section provides foundation and fundamental concepts within the field of occupational S&H.

FUNDAMENTAL CONCEPTS

Incident Prevention

The primary purpose of S&H in the workplace is to prevent incidents and to protect people from injuries and illnesses. Notice the use of the word "incidents" and not "accidents." An accident implies a loss occurred immediately after a short, one-time event. It also implies the event was unpreventable and unpredictable. Conversely, an incident may or may not result in a loss. An incident could be an injury, an illness, damage to equipment, or may be something that nearly resulted in a loss (a near-miss). Furthermore, an incident may not be a short, one-time event. For example, hearing loss occurs over time after continuous, multiple instances of exposure to high levels of noise. For these reasons, it's best practice to use the word incidents and not accidents. As S&H professionals, we should promote that incidents are predictable and preventable. If hazards are adequately anticipated and controlled ahead of time, then incidents can be prevented.

Incidents are most often caused by a series of events or several independent or interacting hazards that contribute to the incident. We refer to all the hazards that lead to an incident as contributing causes, because they each play a part in contributing to the injury, illness, or loss. These hazards or contributing causes should be addressed immediately, and we usually refer to these immediate actions as corrective actions. These actions correct problems right away, such as removing a hazard that caused a trip or guarding a moving part that caused an amputation.

Beyond immediate causes and corrective actions, incident prevention should go step further and ask: "What caused the contributing causes?" We call the answers to this question the root causes, because they are the deeper root of the problem. Root causes should always be linked to high-level S&H program gaps. S&H programs are management systems or policies that direct employees in a way to maintain S&H. Typically, root causes are latent (hidden) failures or weak links in organizational systems, policies, or programs. These root causes lead to contributing causes of incidents. For example, the organization failed to implement a periodic inspection program (root cause) to identify and correct trip hazards (contributing cause). As another example, the organization failed to implement a machine guarding program (root cause) to abate hazards related to moving parts and unguarded machinery (contributing cause). Admittedly, the concepts of immediate corrective actions and proactive preventative actions can have some overlap. But, the solutions to root causes are typically referred to as preventative actions, because they are more long-term solutions that prevent the reoccurrence of similar incidents in the future.

Ideally, incident prevention is achieved by companies implementing proactive incident prevention programs. Some programs (i.e., plans, collection of processes, strategies, usually written down) are required by OSHA standards based on the presence of specific hazards in the workplace such as a Respiratory Protection Program or a permit-required confined space program. When incidents do happen, they should be investigated to implement corrective and preventative measures. Incident investigation is a broad topic in itself. This book does not cover the methods (one might say, art) of incident investigation; however, investigation is an essential practice of S&H professionals. While proactive incident prevention via the implementation of preventative S&H programs is the preferred approach, the reactive incident investigation is also necessary. Both approaches help keep employees safe, but what does "safe" mean?

We define safe as being relatively free from harm. We use the adverb "relatively" because no one, ever, is totally and completely free from harm. At any moment "acts of God" or extremely unlikely events could cause you harm. Also, there is some amount of risk from hazards involved in everything you do. For example, every day you drive your car, you are at risk and not completely safe. But, if you obey the speed limit, practice defensive driving, and do other things to control hazards and reduce your risk, then one might say you are practicing "safe" driving; however, you still could be a victim of a crash perhaps due to unsafe actions of other drivers. At work, we cannot put employees in a bubble and say they are totally safe. There are limitations related to feasibility, money, resources, and allowing people to do their job effectively and efficiently. We have to be practical while following the minimum requirements outlined in occupation S&H regulations. Employers must control hazards likely to cause harm and reduce risk to an acceptable level such that the workplace is relatively free from recognized hazards. This concept summarizes the general duty employers have under OSHA.

Occupational Safety and Health Hazards

The primary function by which the field of occupational S&H prevents incidents is by anticipating, identifying, evaluating, and controlling hazards. Put plainly, hazards are anything that can cause harm. Hazards are anticipated by considering safety during the design and planning of products, processes, and so on. Hazards not only need to be considered during the planning stages but should be identified during real-time operations and production. When hazards are identified, they need to be evaluated for the risk they pose. Not all hazards are created equal. Hazards are evaluated by assessing the risk they pose. Risk considers both the likelihood that a hazard will cause harm and the severity of the consequences the hazard may bring. Once hazards are identified and determined to pose a risk that should be addressed, they should be controlled. Controlling a hazard means taking action to eliminate the hazard or taking action that reduces the risk that a hazard poses. The primary function of S&H in the workplace can be boiled down to recognizing hazards (proactively and in real time), evaluating hazards (assessing risk), and then controlling hazards (eliminating hazards or reducing risk).

Hazard Recognition

Recognizing a hazard means anticipating and identifying anything that can cause harm or that poses a threat. The definition of hazard may seem vague, but there are several approaches to help clarify. Hazards can generally be considered as either unsafe acts or unsafe conditions. Hazards can be grouped into descriptive categories, or hazards can be thought of as how an injury occurs (e.g., mechanism of injury). To sufficiently anticipate and recognize hazards, they should be considered using all approaches.

Usually, hazards can be described as either unsafe acts or unsafe conditions. Unsafe acts are things people do or things they do NOT do. Unsafe conditions are physical characteristics of the environment or of something tangible. Usually, incidents are caused by a combination of both unsafe acts and conditions. Table 1.1 identifies potential unsafe acts and unsafe conditions.

Table 1.1   Examples of Unsafe Acts and Conditions

Hand Cut by Rotating Blade of Machine		*Foot Crushed by Forklift*	
Unsafe Acts	*Unsafe Conditions*	*Unsafe Acts*	*Unsafe Conditions*
• Using incorrect operating procedure • Failure to inspect equipment	• Lack of machine guard • Faulty machine hardware	• Operating machine without training • Failure to use horn when backing up	• Backup alarm not working • Work environment forces pedestrians to cross forklift traffic

Besides unsafe acts and conditions, another way to consider hazards is by placing them into categories. Some occupational S&H organizations categorize hazards as chemical and dust hazards, biological hazards, ergonomic hazards, safety hazards, physical hazards, and organizational hazards. Examples are provided below.

- Chemical and dust hazards
  - Examples: hazardous air contaminants, toxic gases, corrosive materials, and dust particles that are harmful to inhale
- Biological hazards
  - Examples: hepatitis B in spilled blood, rabbis in the saliva of a wild animal, and allergens from peanut oil
- Ergonomic hazards
  - Examples: muscle strains from lifting heavy objects and cumulative trauma disorders from repetitive motion (inflammation of tendons or ligaments)
- Safety hazards
  - Examples: falling from elevated heights, slips, trips, electric shock, cuts, and amputations
- Physical hazards of the environment
  - Examples: radiation, extreme temperature, noise, and vibration
- Organizational hazards
  - Examples: workplace violence, psychological trauma, and emotional stress

Beyond categorization, another way to consider the types of hazards for recognition is by thinking about how people can get hurt or become ill. For physical hazards, these can be mechanisms of injury or how an incident occurs. For health hazards, these can be the cause of disease or the path of a harmful substance into the body. Some examples include:

- Struck by object
- Stuck body part against object
- Slip and fall
- Sprain or strain from lifting
- Fall from elevations
- Electric shock
- Cuts, abrasions (scrapes), and punctures
- Caught in or between
- Fire and explosion
- Inhalation of vapors, fumes, dusts, etc.
- Chemical absorption through the skin
- Ingestion of toxic substance
- Injection of a contaminated substance

- Hearing loss
- Heat exhaustion

The list of examples for hazard exposure and mechanisms of injuries and illnesses could be quite extensive. We can also consider health hazards by classifying chemicals by their effects. Chemical classifications include: corrosive, toxic, carcinogen, irritant, and sensitizing. After anticipating, identifying, or recognizing hazards, they should be assessed to determine the risk they present.

Hazard Evaluation and Risk Assessment

Once hazards are identified, they typically need to be evaluated to know what degree of risk they pose. Risk is a product of the rating of probability a hazard will cause harm and the rating of severity of harm resulting from a hazard. Not all hazards are considered equally. Some hazards may only result in minor injuries or incidents. Others may result in catastrophe. Some hazards may be very unlikely to cause harm. Others may be very likely or frequently present in the workplace. Hazards that present the highest risk are those that are both very likely and can potentially result in catastrophe. Hazards that present the least risk are those that are both very unlikely (infrequently present) and will only result in outcomes of negligible (very low) severity. Both factors (probability and severity) can be rated qualitatively or quantitatively. If using a quantified rating for probability and severity, the number ratings assigned to both probability and severity can be multiplied to calculate risk. It is up to the evaluator to determine a frame of reference for the ratings, assess the ratings, and determine risk. Figure 1.1 provides a risk matrix that can be used as a tool to assess risk. Figure 1.2 is a modified matrix provided by the author to help in assessing each risk factor and provides a reference for ratings of risk factors. The evaluation of hazards and risk assessment allows for employers to prioritize hazard mitigation. The highest risk hazards should of course be addressed first and allows for the proper allocation of time and money to make the biggest impact on risk reduction. Ultimately, it is up to the employer's management to assess each hazard and specific situation and determine what level or rating of risk is acceptable based on the circumstances. S&H professionals are hired to help anticipate and identify hazards, provide recommendations relative to risk assessment and the acceptable level of risk, and provide advice on hazard control methods. It should never be completely and individually up to a S&H professional to make a final decision on an acceptable level of risk or how "safe" is "safe enough." Instead, it is up to the collective decision of management to decide. A S&H professional should be hired to be a topic expert, advise, and recommend.

		Consequence				
		Negligible 1	Minor 2	Moderate 3	Major 4	Catastrophic 5
Likelihood	5 Almost certain	Moderate 5	High 10	Extreme 15	Extreme 20	Extreme 25
	4 Likely	Moderate 4	High 8	High 12	Extreme 16	Extreme 20
	3 Possible	Low 3	Moderate 6	High 9	High 12	Extreme 15
	2 Unlikely	Low 2	Moderate 4	Moderate 6	High 8	High 10
	1 Rare	Low 1	Low 2	Low 3	Moderate 4	Moderate 5

Figure 1.1

To be clear, this summary of risk assessment only provides a very basic overview. There are whole textbooks devoted to risk management. This section provides the most fundamental concepts for a young S&H professional relative to hazard evaluation and risk assessment.

Hazard Control

After hazards are identified and evaluated, hazards should be controlled to prevent incidents, injuries, and health effects. To control a hazard means to take corrective or preventative action in a way that eliminates the hazard or reduces the risk it poses. As we learned in the evaluation section, to reduce the risk of a hazard means reducing the probability the hazard will harm people or reducing the severity of the harm that can result. The most fundamental concept of occupational S&H and incident prevention is understanding and applying the hierarchy of controls.

The hierarchy of controls is a widely accepted framework for reducing risk in the workplace. OSHA and many S&H organizations promote the use of the hierarchy of controls to address hazards effectively. The preferred approach to address a hazard is of course to eliminate it. If you cannot eliminate the hazard altogether, you should then consider ways you can substitute the hazard for something harmless or apply engineering means to physically prevent

Probability of Injury	Severity of Injury			
	4 Catastrophic	3 Critical	2 Marginal	1 Negligible
	Death, Permanent disability, Hospitalization or catastrophic property destruction	Serious injury can result in medical treatment and more than a couple days of loss work time or critical property damage (major repairs)	Major first aid or treatment beyond first aid, but does not result in more than a couple days of loss work time or marginal property damage (minor repairs)	Minor injury, first aid treatment or less or negligible property damage (no repairs)
Frequent 5 — Likely to occur	20	15	10	5
Probable 4 — Likely to occur several times or from time to time	16	12	8	4
Occasional 3 — Likely to occur at sometime	12	9	6	3
Remote 2 — Unlikely to occur, but possible	8	6	4	2
Improbable 1 — So unlikely, Practically assume it will not happen	4	3	2	1

Figure 1.2

people from being harmed. If those preferred methods do not work, employers should then consider applying administrative controls in the form of safe working procedures. The last line of defense for hazard control or protection is the use of personal protective equipment (PPE). Figure 1.3 illustrates the hierarchy of controls.

Elimination is always the preferred method of hazard control. This is because, if you eliminate the hazard, you eliminate the risk altogether. For example, if an entire process or a machine is removed from a factory, any and all hazards from that process or machine are now removed and pose no threat to employees. If you remove a chemical from an industrial facility, the probability of it harming someone is zero; and therefore, the risk of exposure is zero. While elimination is the best option, as you can imagine, it is not always feasible to completely remove a process, equipment, chemical, or other hazard. After all, a business needs to operate and make money and cannot just remove all the things that help them produce products and meet their bottom line.

If you cannot completely remove a hazard, process, or piece of equipment, perhaps you can substitute in something that is non-hazardous. For example, let's say a company uses a flammable liquid to strip paint off of parts. Let's narrow the focus on the flammability hazard. Perhaps, we can

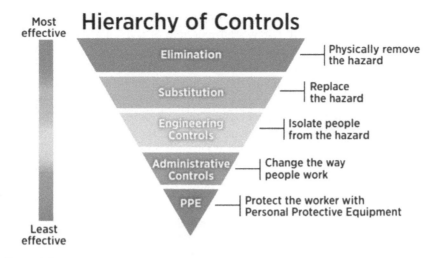

Figure 1.3

substitute out the flammable liquid and substitute in a stripping agent that is not flammable. Companies must be careful with this substitution control method. Substitution is typically only effective when focusing on a specific hazard and may introduce new hazards. For example, perhaps the new paint stripping chemical is not flammable but is toxic instead. As a similar example, let's say we remove the paint stripping chemical and substitute in a process of physically grinding the paint off the part. Well, now we have to worry about paint dust in the air. Substitution is tricky, and it should be used with caution. On another note, if you do successfully substitute out a hazard and substitute in something not hazardous, one can argue this is a form of elimination. That said, effective substitution can blend with elimination. If used carefully, substitution can be effective for controlling hazards and reducing risk.

After consideration of eliminating or replacing a hazard, another good option to reduce risk is by using engineering controls. These engineering controls are tangible things (usual forms of technology or hardware) that typically isolate the hazard from people or isolate people from the hazard. Most importantly, engineering controls do not rely on the proper behavior of people (for the most part). They work no matter what a person does, as long as they don't blatantly, purposefully defeat them. Another way to think of engineering controls is safety by design. Often, engineering controls need to be considered proactively when machines, equipment, and processes are being designed, because these are usually safety devices and countermeasures that are built-in or part of the system. For example, most machine guards would be considered engineering controls. Guards that enclose moving parts provide built-in protection. These fixed guards isolate the hazard from people rather than relying on people to avoid the hazard. Figure 1.4 shows an example of a fixed guard over a belt and pulley system on a machine. An interlocked guard also is safety by design. An interlocked guard isolates a hazard. If the interlock is opened, an electric circuit is opened, and this shuts off the machine and puts it in a safe state while the hazard is exposed. An interlocked guard is another type of engineering control that does not rely on human action to work properly. Figure 1.5 shows an interlocked guard on a lathe (a machine that turns a material to be cut by a cutting head). A self-adjusting guard can be seen in figure 1.6. A self-adjusting guard usually protects the point of operation of a machine like the blade of a table saw. The self-adjusting guard of a table saw automatically adjusts based

Figure 1.4

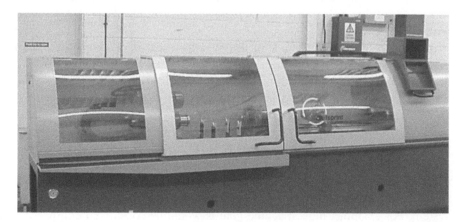

Figure 1.5

Figure 1.6

on the size of the wood. This provides safety design without the reliance on human action and procedures.

Engineering controls are not just machine guards. Engineering controls are methods that reduce the risk that do not rely on human action. Another example of an engineering control we are seeing more and more in vehicles is automatic emergency braking (AEB). AEB uses sensors on the front of the vehicle which detect when there is about to be a crash. If parameters indicate a crash is likely, the vehicle automatically applies the brakes usually much faster than a human can to hopefully avoid the crash. Rather than relying on a human to avoid the hazard, AEB uses engineering means and technology to avoid the hazard. It is important to remember the key characteristic of an engineering control is that it does not rely on human action. For example, some vehicles are now equipped with blind spot warning systems that issue an audible or visual warning if another car is in your blind spot. While this is a hazard control method that uses engineered technology to reduce the probability of a hazard (crash), it is a warning device that relies on proper human action. Therefore, a blind spot warning system is not an engineering control but an administrative control. Another simple example of an engineering control is a workroom or enclosure that totally isolates an employee from a hazardous environment. A final example of an engineering control is ventilation. Adding fresh air to the workplace (dilution ventilation) can increase the volume of clean air and thus reduce the concentration of contaminants. Local exhaust ventilation removes hazardous air contaminants at the source where they are produced. Local exhaust ventilation is another engineering control for airborne health hazards and an example is shown in figure 1.7 as applied to a welding operation. Ventilation is an engineering control because once it is active, it reduces the risk of air contaminants and does not rely on human action. Ventilation is a common engineering control for hazardous air contaminants.

Figure 1.7

Engineering controls are a preferred approach to risk reduction, because they do not rely on human behavior. Humans make mistakes. Humans get distracted. Humans cut corners. Humans are not perfect. Engineering controls, machine guards, and advanced technologies that do not rely on the imperfect human are effective ways to control hazards and reduce risk.

If hazards cannot be eliminated, replaced, or controlled by engineering means, then employers should use administrative controls. Unlike engineering controls that do not rely on human action to be effective, administrative controls are procedures that rely fully on employee execution and compliance. Administrative controls are protocols and procedures. Examples of administrative controls are easy to come up with, because they are anything that reduces risk or controls a hazard that relies on human action. The following are examples of administrative controls:

- Equipment pre-use or regular inspections to identify hazards
- Warning signs and signals that tell employees to do something (e.g., alarms and blind spot warning signals in vehicles)
- Protocols that direct employees to do or not do something (e.g., safe operating procedures)
- "DO NOT ENTER" and "STOP" signs, or any warning or directive signs
- Caution tape
- Procedures that rotate employees in departments to reduce exposure to noise
- Procedures that limit employees from being in a department for longer than a certain period of time to reduce exposure to a hazardous air contaminant
- A work permit system that requires supervisor sign-off to authorize work
- Training
- Supervision

Because administrative controls rely on human action, and because humans naturally make mistakes, they normally should be layered with other controls. For example, even if a machine has engineering controls like machine guards, employees should still be trained on proper use and have procedures on how to maintain the machine in safe operating conditions. When engineering controls are not feasible and hazard control totally relies on administrative controls, then administrative controls should be layered together to adequately reduce risk. Some S&H professionals may consider PPE a subset of administrative controls.

PPE are articles that are worn on an employee's body to protect them from hazards. Figure 1.8 illustrates common PPE. PPE is the least preferred method of control; however, it is a common and necessary method of control. PPE is often referred to as the last line of defense for hazard control. This is because PPE does not directly affect the hazard at all. Instead, PPE simply reduces the severity of the hazard if it does reach an employee's body. PPE reduces the severity, but it does not decrease the probability of harm. PPE should always be layered with higher-level controls. One may consider PPE a subset of administrative controls, because if employees are

Head protection
Falling or flying objects, overhead objects

Eye protection
Blowing dust or particles, metal shavings, acids or caustic liquids, welding light

Hearing protection
Loud tools and machinery, poorly maintained equipment

High-visibility hat, vest, pants
Errant vehicles, distracted drivers

Chaps pants
Chainsaws

Hand protection
Sharp or hot objects, chemicals, biological or electrical hazards

Foot protection
Falling or rolling objects, sharp or heavy objects, wet and slippery surfaces, uneven surfaces, hot surfaces, electrical hazards

Harness lanyard
Working more than 6 feet or more above a lower level

Figure 1.8

required to wear PPE, then there must be an established policy that directs employees to wear PPE. Further, employees must be trained on the proper use and limitations of PPE. While PPE is common, it should be considered the last line of defense in a layered approach to hazard control and incident prevention.

The primary purpose of occupational S&H is to prevent incidents, injuries, and illnesses. The main function by which this purpose is achieved is by identifying, evaluating, and controlling hazards. Hazards should be identified proactively and during real-time operations. Hazards are evaluated by assessing the probability of harm and severity of harm in order to determine risk. Risk allows us to prioritize hazards for control. We control hazards by applying the hierarchy of controls. Companies should first always consider methods to eliminate or replace a hazard. If a hazard cannot be eliminated or replaced, proactive engineering controls should be applied that isolate the hazard from people without relying on human action. Finally, in lieu of engineering controls or in addition to engineering controls, administrative controls or procedures should be layered to reduce risk. PPE should be used as a last line of defense and to supplement other controls within the hierarchy. These fundamental concepts are the foundation to understanding incident prevention programs and OSHA regulations.

REVIEW QUESTIONS

1. Describe the general storyline of occupational S&H from the early 1800s to the end of the 1900s.
2. What's the difference between an accident and an incident as defined in this chapter? Are all incidents preventable? Why or why not?
3. What is the primary purpose of the field of occupational S&H, and how do employers achieve this?
4. What is a hazard? Give five examples. Describe two approaches for considering hazards to aid in hazard identification.
5. What is the difference between a hazard and risk? How do we assess hazards and calculate risk? What advantage is there for assessing the risk of hazards?
6. Describe the hierarchy of controls in order of methods. Give an example of each method of control.
7. What are the two limitations to PPE?

Chapter 2

The Occupational Safety and Health Act and the Occupational Safety and Health Administration

INTRODUCTION AND SCOPE

The Occupational Safety and Health Act (OSH Act) was signed into law in 1970 to protect the health and safety of American workers. The OSH Act applies to practically all private employers in the United States. The act establishes responsibilities for both employers and employees regarding health and safety. The primary responsibility of employers, under the OSH Act, is to provide a workplace relatively free from recognized hazards. The OSH Act also provides rights for both employers and employees relative to safety and health. The OSH Act is supported and enforced by the Occupational Safety and Health Administration (OSHA).

OSHA provides education, support, training, and enforcement for the OSH Act and standards promulgated under the act. OSHA standards are found in Title 29 of the Code of Federal Regulations (CFR). Standards of Title 29 are separated into parts by industries to which they apply (e.g., General Industry, Construction). OSHA standards are enforced during OSHA inspections conducted by compliance officers. Inspections are either programmed (planned) or unprogrammed (unplanned). OSHA prioritizes inspection types based on hazards involved. OSHA inspections almost always include a physical walk around the workplace as well as a review of health and safety records.

As a result of OSHA inspections, employers can receive citations (official government notice) for observed violations of OSHA standards. A citation may or may not have a monetary penalty associated with it. The size of the penalty depends on the degree of the violation or severity of the hazards involved. Employers can contest citations. But most commonly, employers will abate (correct, fix) the hazard within the required abatement period and respond appropriately to the citation.

In addition to enforcement, OSHA provides safety and health training and cooperative programs. OSHA's website has many resources to help employees and employers with hazard identification and control. OSHA facilitates the OSHA Training Institute devoted to providing safety and health education and training to employees, employers, and safety professionals. OSHA also facilitates compliance assistance programs and partnerships that help employers come into compliance and prevent workplace incidents.

This chapter covers the OSH Act and topics related to the OSHA regulatory and enforcement process.

THE OSH ACT OF 1970

In 1970, the US Congress passed the OSH Act of 1970. President Nixon signed the act that year, and it became US law. OSHA was established to support and enforce the act and to provide workers safety and health protection. The act was passed for two primary purposes. For one, the OSH Act was intended to ensure the United States remained economically competitive. Congress found personal injuries and illnesses arising out of work imposed a substantial burden upon, and hindrance to, US production. The second reason was to preserve human resources and save lives. The following methods are presented in the OSHA Act regarding how the US Congress will meet these purposes:

- By encouraging employers to reduce safety and health hazards in the workplace
- By providing employers and employees separate responsibilities and rights with respect to achieving safe and healthful working conditions
- By creating an Occupational Safety and Health Review Commotions for carrying our adjudicatory (legal review) functions under the act
- By providing research in the fields of occupational safety and health and developing innovative methods for dealing with safety and health problems
- By exploring ways to discover latent causes of disease and establishing causal connections between the work environment and health problems
- By providing training programs to increase the number and competency of personnel engaged in the field of occupational health and safety
- By providing for the development of occupational safety and health standards
- By providing an enforcement and inspection program to enforce safety and health standards
- By providing appropriate reporting procedures with respect to health and safety
- By encouraging joint labor-management efforts to reduce injuries and illnesses arising out of employment

Applicability of the OSH Act

The OSH Act (and therefore federal OSHA enforcement) applies to all private-sector employers in all fifty states unless otherwise covered by an OSHA-approved State Plan. Employers not covered under the OSH Act include:

- Self-employed workers
- State and local government workers
  - Exception: The US Postal Service is covered by the OSH Act and federal OSHA
- Family farms that only employ family members
- Workplace hazards regulated by another federal agency (e.g., Mine Safety and Health Administration, the Department of Transportation, Coast Guard)

State Programs

The OSH Act allows states to create their own occupational safety and health plans and administrations. State Plans are approved by federal OSHA and operated by individual states or US territories. State Plans are monitored by federal OSHA and must be at least as effective as OSHA in preventing workplace injuries and illnesses. This means the laws enforced by State Plans must be at least as strict as federal OSHA regulations. Some State Plans only cover state/local government workplaces (not covered by federal OSHA), while other State Plans cover both private (nongovernment) employers and state/local government employers. Figure 2.1 illustrates states that have state OSH plans for state-level enforcement and support of the OSH Act.

Responsibilities and Rights under the OSH Act

Employer Responsibilities and Rights

The OSH Act of 1970 explicitly describes the primary duty of employers under the OSH Act. This is referred to as the General Duty Clause (OSH Act Section 5(a)(1) and (2)):

5(a) Each employer—

(1) shall furnish to each of his employees employment and a place of employment which are free from recognized hazards that are causing or are likely to cause death or serious physical harm to his employees;

(2) shall comply with occupational safety and health standards promulgated under this Act

This state's OSHA-approved State Plan covers private and state/local government workplaces.

This state's OSHA-approved State Plan covers state/local government workers only.

This state (with no asterisk *) is a federal OSHA state.

Figure 2.1

The General Duty Clause is important, because it summarizes the overall responsibility of employers under the OSH Act. It is also important to know that if there is no specific OSHA standard which covers a hazard that poses an unacceptable risk, OSHA can cite the hazard using the General Duty Clause instead of a specific standard. For example, there is not a federal OSHA regulation or standard that covers hazards of repetitive motion in the workplace. But, if an employer subjects employees to work that involves repetitive motion, and employees are consistently developing musculoskeletal disorders (e.g., Carpal Tunnel) from that repetitive motion, then OSHA can cite the hazard using the General Duty Clause to penalize the employer for failing to control the recognized hazard.

Beyond employers' general duty under the OSH Act, the act and OSHA outline the general responsibilities of employers. The following list summarizes some of the employer requirements under the OSH Act:

• Provide a relatively safe workplace free from recognized hazards
• Comply with OSHA standards
• Evaluate the workplace for hazards

- Report deaths and serious injuries to OSHA
- Provide safety training to employees
- Comply with OSHA required posting (e.g., citations)
- Maintain OSHA required recordkeeping of injuries and illnesses
- Provide safe tools and machinery to employees
- Cooperate with OSHA inspection protocols and correct OSHA violations

One specific responsibility to note is employers must post the OSHA "It's the Law" poster in a prominent location in physical workplaces. Figure 2.2 shows the OSHA poster.

Employers also have rights under the OSH Act and OSHA. Major employer rights are outlined below. Employers have the right to:

Figure 2.2

- Seek free advice from OSHA
- Request and receive proper identification of OSHA compliance officers
- Have an opening and closing conference with the compliance officer
- Accompany an OSHA compliance officer on an inspection
- Dispute an OSHA citation/violation
- Apply for a variance from a standard's requirements
- Take an active role in developing safety and health programs
- Submit information or comments to OSHA on the issuance, modification, or revocation of OSHA standards and request a public hearing

Employee Responsibilities and Rights

Employers are mostly responsible for safety and health in the workplace, but employees have responsibilities under the OSH Act and OSHA as well. Some major employee responsibilities are summarized below. Employees should:

- Read the OSHA "It's the Law" poster at the jobsite
- Comply with all applicable OSHA standards
- Follow all employer safety and health rules and regulations, and wear or use prescribed protective equipment as needed
- Report hazardous conditions to the supervisor
- Report any job-related injury or illness to the employer, and seek treatment promptly
- Cooperate with the OSHA compliance officers
- Exercise your rights under the OSH Act in a responsible manner

In addition to outlining employee responsibilities, the OSH Act and OSHA provide employee rights. Some rights are summarized below. Employees have the right to:

- Review copies of appropriate OSHA standards, rules, regulations, and requirements
- Request information from your employer on safety and health hazards, precautions, and emergency procedures
- Receive adequate training and information
- Request that OSHA investigate if you believe hazardous conditions exist
- File an anonymous complaint with OSHA
- Respond to questions from OSHA compliance officer
- Be involved in an OSHA inspection and have questions answered
- Review records of workplace injuries and illnesses
- Submit information or comments to OSHA on the issuance, modification, or revocation of OSHA standards and request a public hearing

OSHA

OSHA was created to support the provisions of the OSH Act. Since the creation of OSHA, the worker injury, illness, and fatality rate have been reduced significantly. OSHA conducts a wide range of programs to promote health and safety. OSHA's primary responsibilities include enforcement of the OSH Act and related health and safety standards, outreach and education regarding occupation safety and health, and cooperative programs with industry to reduce workplace incidents.

OSHA is part of the executive branch of the US government. More specifically, OSHA is an agency within the Department of Labor. The OSHA administrator (head, director) is the assistant secretary of labor for occupational safety and health who reports to the secretary of labor. The secretary of labor is a member of the president's executive cabinet.

OSHA REGULATIONS AND STANDARDS

OSHA standards under the OSH Act are regulatory laws that employers must follow. The OSH Act is Title 29 of the Code of Federal Regulations (CFR). Title 29 is the title of the CFR pertaining to labor and is separated into parts. The following parts of Title 29 of the CFR fall under OSHA jurisdiction:

- Recordkeeping Standards (Part 1904)
- General Industry Standards (Part 1910)
- Maritime Standards (Parts 1915, 1917, and 1918)
- Construction Standards (Part 1926)
- Agriculture Standards (Part 1928)

This book exclusively summarizes compliance with Part 1910 - General Industry Standards (those not falling under specific industries of other parts). Each part of Title 29 has lettered subparts (e.g., A, B, C, D). As an example, let's look at a standard in Subpart D of the OSHA standards for General Industry referring to requirements for walking and working surfaces.

An OSHA standard (requirement) is written (cited) as follows:

- 29 CFR 1910.25(b)(3)
  - 29 CFR: Title 29 of the CFR pertaining to labor
    - 1910: Part of the code-title applicable to General Industry
      - 25: section number of a subpart (in this case subpart D, subparts are not included in citation)
        - (b): major paragraph
          - (3): paragraph subsection
            - And further . . .

OSHA regulations and standards can be created, updated, amended, or revoked by OSHA.

Creating or Changing an OSHA Standard

To create, adopt, change, or revoke an OSHA standard, the secretary of labor must receive a base of information and reasoning submitted to him in writing. This base of information can come from other government agencies and departments related to safety and health (e.g., NIOSH – more on this agency later), state/local governments, employer representatives, nationally recognized standard producing organizations, or other interested parties). The secretary shall then provide an OSHA advisory committee with his/her recommendation regarding the new proposed ruling (standard) or change. This recommendation is developed with input from other government agencies related to occupational safety and health and subject experts regarding the proposal. OSHA has several standing advisory committees that advise the agency on safety and health issues.
   The two standing advisory committees are:

• The National Advisory Committee on Occupational Safety and Health (NACOSH), which advises OSHA on the administration of the OSH Act
• The Advisory Committee on Construction Safety and Health (ACCSH), which advises on issues specific to the construction industry

Other continuing advisory committees include:

• The Federal Safety and Health Advisory Committee (FACOSH), which advises specifically on federal-level issues
• The Maritime Advisory Committee for Occupational Safety and Health (MACOSH), which advises on issues specific to the maritime industry
• The National Advisory Committee on Ergonomics, which advises on initiatives regarding ergonomics (e.g., repetitive motion and heavy lifting)

   After the secretary of labor's recommendation is reviewed by an advisory committee, it can move forward. Generally, the advisory committee has ninety days to provide its recommendation back to the secretary of labor. OSHA then shall publish the proposed rule, modification, or revoking of a standard in the Federal Register. This is done as either a Request for Information (soliciting information to be used in drafting a proposed rule) or a Notice of Proposed Rulemaking (proposed new rule's requirements). After this, there is a commenting period for people, employers, and interested parties to provide written comments or objections. OSHA may schedule a public court hearing to consider various viewpoints. After the commenting period, OSHA publishes the full text of a standard with explanation and justifications supporting it or a determination that no addition or change is necessary. OSHA submits these final rulings to Congress. Congress has the authority to pass or repeal the rule. Final rulings or new standards are added to the CFRs.

Variances from Standards or Regulations

Although rare, employers can apply for variances from OSHA standards or regulations, and OSHA may approve such variances. A variance can be temporary, permanent, or experimental. A temporary variance means the employer claims they cannot comply with a standard or new rule by its effective date because of a lack of technical personnel, material, or equipment, or because of necessary construction of physical facilities at the time. When operating under a temporary variance, the employer must meet certain conditions specified by OSHA for the time period granted. An employer may apply for a permanent variance if they can prove working conditions, practices, means, methods, operations, or processes at the worksite are as safe and healthy as they would be if they complied with the standard. An employer may also apply for an experimental variance. If granted, experimental variance allows for the employer to demonstrate or validate that new hazard control techniques are effective. Variances are not common; however, they do exist.

OSHA INSPECTIONS

While OSHA conducts a broad range of programs and activities to protect the safety and health of employees, one of their main strategies is strong, fair, and effective enforcement of standards and regulations. OSHA enforces standards primarily by performing workplace inspections. OSHA inspections can be classified as programmed or unprogrammed. Programmed inspections are planned by OSHA in advance. These inspections might be for emphasis programs for high-hazard employers or may be enhanced enforcement programs for targeted employers with poor safety performance. OSHA may conduct an inspection with advanced notice to the employer for situations where the inspection must occur after regular business hours, might require specific employees or representatives to be present, or might be more productive with advanced notice. Unprogrammed inspections are not scheduled in advance. These unprogrammed inspections may be done at random, in response to employee's complaints, in response to observed hazards that are immediately dangerous, or as a result of a reported incident. Generally, inspections are done without prior notice. Anyone who alerts an employer in advance of an unprogrammed inspection can be fined up to $1,000 or receive a six-month jail term or both. When an OSHA compliance officer shows up, the employer has the right to refuse entry into the workplace; however, the compliance officer can obtain a warrant and come back. OSHA inspections are prioritized based on the hazards that trigger them.

OSHA Inspection Priorities

OSHA cannot inspect the millions of workplaces covered by the OSH Act every year, so the most hazardous workplaces are prioritized for inspection. An inspection system has been established by prioritizing inspection types. Priorities are outlined and described below in order of highest to lowest priority:

1. Imminent danger—a condition is present where there is reasonable certainty that a danger exists that can cause death or serious harm immediately or before the danger can be eliminated by OSHA enforcement procedures; imminent danger situations can be observed by Compliance Officers or can be reported by someone
2. Catastrophes and fatal incidents—an inspection in response to incidents resulting in the death of any employee or hospitalization of one or more employees; this is the second priority behind imminent danger, because the incident and loss unfortunately already occurred
3. Employee complaints—inspections in response to employee complaints that involve serious hazards
4. Planned or programmed inspections—inspections in industries with a high number of hazards, injuries, or illnesses
5. Follow-up—inspections to previous inspection types

OSHA Inspection Process

When an OSHA compliance officer arrives at a workplace for an inspection, they are required to display credentials (identification proving they represent an OSHA area office) and ask to meet with an appropriate employer representative. The OSHA compliance officer will conduct an opening conference with the employer's selected representative. During the opening conference, the office will explain why the inspection is occurring (e.g., result of compliant, random), obtain information about the employer, and explain the scope of the inspection (e.g., physical areas to inspect, employee interviews, and review of records). An employer representative is allowed, but not required, to accompany the compliance officer during the physical inspection (walk-around).

During the inspection walk-around, the compliance officer must be allowed to proceed through the establishment freely to inspect for potentially hazardous working conditions. The walk-around may cover all or part of the physical, and point out OSHA standard violations. The officer will likely discuss possible corrective actions for conditions that violate OSHA standards. The compliance officer may consult, at times privately, with employees during the inspection. Some apparent, easily-corrected violations can be corrected immediately during the walk-around. These immediately corrected violations may still serve as the basis for a citation (of a violation) or penalty.

After a walk-around inspecting physical working conditions, the compliance officer is likely to check for required OSHA postings (e.g., OSHA poster, an annual summary of incidents, past OSHA citations, etc.). The compliance officer may also request records (logs) of injuries and illnesses, as well as training records. The officer will end the visit with a closing conference. During the closing conference, the officer typically discusses any violations for which a citation may be issued and describes that employers have the right to appeal citations and/or corresponding penalties within fifteen working days of receiving a citation. The next section describes OSHA citations, the process of responding to or repealing citations, and penalties for violations.

OSHA CITATIONS AND PENALTIES

After an inspection, the compliance office will complete an inspection report and submit the report to his or her area director (supervisor) for review. After reviewing a compliance officer's inspection report, the area director of the local OSHA office will determine whether to issue a citation and if the citation will have a proposed penalty. The area director has up to six months to issue a final report with citations to the employer. OSHA violations and penalties are classified based on the severity of the violation and described in table 2.1. OSHA penalty amounts are standard based on the severity of violations. These standard amounts increase periodically based on inflation. The issuing area director may adjust the specific monetary amount for penalties based on the circumstances (e.g., size of employer, good faith shown by the employer, severity of hazard, and history of citations).

An OSHA citation may be issued based on evidence of a violation of an OSHA standard observed during an OSHA inspection. If the Area Director confirms a violation has occurred and a citation will be issued, OSHA will

Table 2.1   OSHA Violation Types and Possible Penalties*

Type of Violation	Description	Maximum Penalty Amount (As of Jan 8, 2021)
De Minimis	No citation issued, condition does not meet OSHA standard but does not pose a hazard	No penalty
Other-than-serious	Violations that would not likely result in severe harm, posting requirement violations	$13,653 per violation
Failure to abate	Not fixing hazard within prescribed citation time frame	$13,653 per day beyond the abatement date
Serious	Substantial probability for death or serious physical harm, employer should have known better	$13,653 per violation
Willful	Employer intentionally or knowingly commits	$136,532 per violation, possible jail time
Repeat	Similar to previous violation at same company within last five years	$136,532 per violation

*Penalty amounts may not up to date as they increase to account for inflation.

issue a citation and notification of penalty to the employer. Citations inform the employer and employees of:

- Regulations and standards the employer allegedly violated
- Possible abatement (correction) measures to be taken
- The length of time set for abatement of hazards
- Any proposed penalties

Upon receiving a citation, the employer must physically post the citation in the workplace. The citation must be posted (or a copy of it) at or near the place where each violation occurred to make employees aware of the hazards to which they may be exposed. The OSHA notice of citation must remain posted for three working days or until the hazard is abated, whichever is longer.

When issued a citation and/or penalty, an employer can either agree to fix the problem or contest the citation, penalty, or abatement date. If the employer accepts the citation, they must correct the violation by the date set for abatement (time period given to fix the problem) and must pay the penalty if one is proposed. The employer must send the issuing area director an

Abatement Certification. For some violations, this may be a signed letter certifying the hazard has been abated. For other more serious violations, detailed proof might be required as indicated in the notice of citation.

If the employer does not agree with the citation or penalty, the employer may contest the citation. Before officially contesting the citation, the employer can request an informal conference with the issuing area director to discuss potential disagreements or options. If the employer decides to officially contest the citation in any way, they must submit an official Notice to Contest to the issuing area director within fifteen working days of receiving the citation. There is no formal format for a Notice of Contest, but it must be delivered in writing to the OSHA area director. A proper contest suspends the employer's legal obligation to abate and pay until the item contested has been legally resolved. After filing a Notice to Contest, the contest is officially in litigation. Upon receiving a Notice of Contest, the area director provides the notice to the Occupational Safety and Health Review Commission (OSHRC). This commission will assign a judge to hold a hearing on the case. The judge will either find the contest legally invalid or set a public hearing to evaluate and make a decision whether or not to overturn the citation. The employer and its employees have the right to participate in the hearing. After the judge's decision, if the employer still does not agree, they can appeal to have the case escalated to US court. This escalation is not common.

Besides contesting violations and penalties, employers can also contest prescribed abatement time periods for violations. If an employer cannot correct the hazard within the required abatement time, they can file a Petition for Modification of Abatement (PMA). A PMA must be submitted to the issuing area director no later than one working day after the originally prescribed abatement date. The PMA must be submitted in writing and include steps you have taken to achieve compliance, how much time is needed, why the employer needs more time, and certification the petition has been posted in the workplace. The area director may grant or oppose the PMA. If the area director opposes the PMA, it becomes a contested case and is sent to the OSHRC to start the litigation process.

OSHA OUTREACH, EDUCATION, AND COMPLIANCE ASSISTANCE

Beyond enforcement, OSHA provides outreach, education, and compliance assistance to employers. On their website, OSHA provides interactive tools and information on hazard recognition, specific topics, and hazard control measures. OSHA also has an extensive publications program which includes numerous publications about specific regulatory topics, compliance guidance, and other safety and health information. OSHA also administers the OSHA Training Institute (OTI). The OTI provides education opportunities for employers, employees, and safety professionals. The OSHA Outreach Training Program is the OTI's primary way to train workers and designate professionals as Outreach Trainers. This program includes OSHA's ten-hour and thirty-hour courses in Construction and General Industry Standards. Professionals can become a designated Outreach Trainer and teach these ten- and thirty-hour courses to employees themselves. Finally, OSHA offers compliance assistance and cooperative programs. Consultation helps employers identify and correct workplace hazards. Small employers are the main target of these consultation services. OSHA also offers cooperative programs. These programs are alternatives to traditional OSHA enforcement strategies. The primary cooperative program is the Voluntary Protection Program (VPP). This allows OSHA to visit the employer's establishments regularly to help with compliance without risk of citation. To be an OSHA VPP workplace, the employer must meet certain criteria and establish an effective safety management system as defined by OSHA for preventing incidents. Although there is a cost of resources for employers to be part of a VPP, there are some real benefits such as avoiding penalties, promoting compliance, and reducing workplace injuries and illnesses. OSHA offers other partnership and cooperative programs as well, but they are not covered in detail in this book.

REVIEW QUESTIONS

1. What is the main purpose of the OSH Act? What is the primary responsibility of employers under the OSH Act? What are five examples of employee rights under the OSH Act?
2. What kind of employers does the OSH Act apply to? What kinds does it not apply to?
3. What are the parts of Title 29 of the CFR? What does Title 29 apply to?
4. What is the process for creating an OSHA standard?
5. What are the types and priorities of OSHA inspections?
6. Describe the OSHA inspection processes.
7. Describe the process of receiving a citation? What is the process to contest a violation?
8. Describe different types of OSHA violations and the penalty for each. Look up current penalty amounts on osha.gov as they increase annually.
9. What else does OSHA do other than enforce standards by conducting inspections?

REFERENCES

U.S. Department of Labor, Occupational Safety and Health Administration, *All About OSHA* (OSHA 3302-01R 2020). https://www.osha.gov/sites/default/files/publications/all_about_OSHA.pdf

————, *Employer Rights and Responsibilities Following a Federal OSHA Inspection.* https://www.osha.gov/Publications/osha3000.pdf

————, *Federal Employer Rights and Responsibilities Following an OSHA Inspection* (1996). https://www.osha.gov/Publications/fedrites.html#:~:text =Posting%20Requirements,is%20abated%2C%20whichever%20is%20longer

————, *OSH Act of 1970.* https://www.osha.gov/laws-regs/oshact/completeoshact

Chapter 3

OSHA Injury and Illness Recordkeeping

INTRODUCTION AND SCOPE

Most employers covered by OSHA, with few exceptions, must keep records of specific types of injuries and illnesses, and all employers covered by OSHA must report certain incidents to OSHA in a timely manner. Those injuries and illnesses that must be recorded are referred to as OSHA-recordable incidents. Those incidents that must be reported directly to OSHA are referred to as OSHA-reportable incidents. Employers can face penalties and OSHA citations if they do not keep injury and illness records properly and report certain incidents to OSHA.

A recordable incident is one that is work-related, is a new case, and meets recording criteria. An incident is work-related if the event or exposure occurred in the work environment or during work duties. An incident is a new case if the employee never experienced the injury or illness before. An incident is also a new case if the employee has experienced the injury or illness before, but he or she has since completely recovered from it and it happens again. To be a recordable incident, the injury or illness must meet recording criteria such as, but not limited to, resulting in missed work or job restrictions, requiring medical treatment beyond first aid, or resulting in a specific outcome. Employers are required to log all recordable incidents that happen to the employees they pay or employees they supervise on an OSHA 300 Form (log). Employers must summarize OSHA 300 Logs annually using an OSHA 300-A Form. Employers are required to document the reporting of incidents, injuries, and illnesses when they happen (e.g., using an OSHA 301 Form). This series of incident recordkeeping forms (OSHA 300 Forms) are required, but equivalent forms can be used. Recordable incident forms (300 and 300-A), for most employers, must be submitted electronically to OSHA each year.

33

A reportable incident is one that results in a death, the in-patient hospitalization of one or more employees, the loss of an eye, or an amputation. Employers must report deaths directly to OSHA within eight hours of the incident. Employers must report all other types of reportable incidents directly to OSHA within twenty-four hours. All employers covered by OSHA must report these types of incidents in a timely manner.

The OSHA standards covered in this chapter include:

- 29 CFR 1904—Recordkeeping:
  ◦ Subpart A—Purpose
  ◦ Subpart B—Scope
  ◦ Subpart C—Recordkeeping Forms and Recordkeeping Criteria
  ◦ Subpart D—Other Injury and Illness Recordkeeping Requirements
  ◦ Subpart E—Reporting Fatality, Injury, and Illnesses to the Government

SUMMARY OF OSHA STANDARDS

Incident Definitions

- Recordable injury or illness (incidents)—an incident that is work-related, is a new case, and meets general or specific OSHA recording criteria that must be recorded on an OSHA 300 Form
- Reportable injury or illness (incidents)—an incident that results in the in-patient hospitalization of one or more employees, an amputation, the loss of an eye, or death that must be reported directly to OSHA

Recordkeeping: Purpose and Scope

The purpose of the OSHA recordkeeping requirements is to require employers to keep records of and report certain work-related incidents and fatalities. All employers covered by the OSH Act are required to keep these records and comply with 29 CFR 1904 recordkeeping requirements with only some exceptions.

Employers with ten or fewer employees do not have to maintain records of injuries and illnesses; however, even these small employers must report reportable incidents to OSHA. This ten-employee exemption for keeping recordable injuries and illnesses refers to the number of employees in the entire company not just the number of employees at a specific company location.

As another exemption example, employers classified as a specific, low-hazard industry group listed in appendix A of 29 CFR 1904 Subpart B are also excused from maintaining records of recordable incidents. Some examples of these low-hazard employer types are: gas stations, clothing stores, florists, radio and television broadcasting stations, bars, and restaurants. These low-hazard employers must still report reportable incidents to OSHA. As a safety and health professional, it is important to know if your company does or does not fall into one of these low-hazard industry groups for exemption.

Recordkeeping Forms and Recording Criteria

Each employer covered under the OSHA Act shall keep records (a log) of fatalities and recordable incidents (if not exempt as described in the previous section). A recordable incident is an injury or illness that is work-related, is a new case, and meets general or specific OSHA recording criteria. The "work-relatedness" and the "new case" parts of the definition are generally straightforward. Before moving on, it is worth first noting the general recordable criteria. To be recordable, an incident must be work-related, be a new case, and meet one or more of the following criteria:

- Result in a fatality
- Result in loss of consciousness
- Result in one or more days away from work, restricted work duties, or transfer to a different job
- Result in medical treatment beyond first aid
- Is a significant injury or illness as diagnosed by a physician or licensed health care professional
- Is a specific type of incident as defined by OSHA (e.g., needle stick, noise-induced hearing loss, etc.)

Proceeding sections provide further detail on how to determine if an incident is work-related, how to identify if an incident is a new case, and how to determine if an incident meets the recording criteria listed above.

Determining Work-relatedness

You must consider an injury or illness to be work-related if an event or exposure in the work environment either caused or contributed to the resulting condition or significantly aggravated a pre-existing injury or illness. Work-relatedness is presumed for injuries and illnesses resulting from events or exposures occurring in the work environment. OSHA defines the work environment as "the establishment and other locations where one or more employees are working or are present as a condition of their employment. The work environment includes not only physical locations, but also the equipment or materials used by the employee during the course of his or her work." So this means, to be work-related, the incident does not have to occur in the building where the employee reports to work. The incident might be work-related and occur while performing job duties outside of their physical work location. While it may be easy to come up with obvious examples of work-related incidents, it is helpful to consider unique incidents that would NOT be considered work-related. OSHA provides examples of when employers are NOT required to record incidents due to non-work-relatedness in 29 CFR 194.5(b)(2). These examples are found in appendix A of this chapter.

Two specific examples of incidents are worth noting when discussing work-relatedness, and those involve incidents of employees who work from home and incidents of employees who are traveling for work. An incident would be work-related for those employees working from home if the employee got injured while performing work for pay in a way such that the injury or illness is directly related to the performance of work rather than to the general home environment or setting. For example, if an employee is injured getting coffee while working from home, the injury is not work-related. If an employee develops and is diagnosed with Carpel Tunnel while working on the computer every day for work, then the incident is work-related. Incidents that occur to employees who are traveling for their job would not be work-related if the incident occurs in their home or away from home (e.g., hotel) or if it occurs when taking a detour for personal reasons. For example, if an employee gets injured while taking a tour of the city during a work trip and that tour was not part of his/her job duties, the injury is not work-related.

Determining If an Incident Is a New Case

An injury or illness is considered a new case if either:

1. The employee has not previously experienced a recordable incident of the same type that affects the same part of the body, or
2. The employee previously experienced a recorded injury or illness of the same type that affected the same part of the body but had recovered completely (all signs and symptoms had disappeared) from the previous injury or illness and an event or exposure in the work environment caused the signs or symptoms to reappear.

There are three clarifications worth noting when it comes to determining if an incident is a new case. If an employee experiences the signs or symptoms (S/S) of a chronic work-related illness, the employer does NOT need to consider EACH reoccurrence of S/S to be a new case. For example, if an employee is diagnosed with silicosis from exposure to silica at work and receives medical treatment, the illness is recorded initially. If the employer does NOT have to record the illness every time the employee develops S/S of the illness and pursues further treatment. The illness is only recorded once. When an employee experiences the S/S of an injury or illness as a result of an event or exposure in the workplace, such as an episode of occupational asthma, then it would be considered a new case since reoccurrence of S/S was caused by an event or exposure in the workplace. Lastly, whenever attempting to determine if a case is a new case or an old case, it is wise to rely on the diagnosis or advice from a physician or licensed health care professional.

Recording Criteria and Forms

General Recording Criteria

To be a recordable injury or illness, an incident must be work-related, be a new case, and meet certain recordable criteria. An incident meets recordable criteria if it results in a death, days away from work, job transfer, restricted duty, medical treatment beyond first aid, or loss of consciousness. An incident also meets recordable criteria if the employee is diagnosed with a significant injury or illness by a physician or licensed health care provider. These are the general recordable criteria.

These recordable criteria can be further explained. "Days away from work" refers to an employee missing work days due to the injury or illness. "Job transfer" means the employee is forced to work a different job to allow for work without further aggravating the injury or illness. "Restricted duty" means the employee's job responsibilities are reduced to allow the employee to continue working without aggravating the injury or illness. "Medical treatment beyond first aid" is best explained by identifying what is meant by first aid. Any medical treatment beyond the measures described in the list below would be considered "medical treatment beyond first aid."

First aid includes:

- Cleaning, flushing, or soaking skin wounds
- Using wound coverings such as Band-Aids and gauze
- Using hot/cold therapy
- Using means of support such as wraps, slings, splints, and back boards
- Draining fluid from nails or blisters
- Using eye patches or coverings
- Removing foreign bodies from the eye using water or a cotton swab
- Removing splinters by simple means
- Massages
- Drinking fluids for relief of heat stress

OSHA provides examples of "significant injury or illnesses" diagnosed by a physician or licensed health care provider:

• Punctured eardrum
• Fractured toe or rib
• Byssinosis, silicosis, cancer
• Chronic irreversible diseases
• Fractured/cracked bones

There are also some additional, specific criteria beyond the general recording criteria that will be discussed later.

OSHA 300 Log of Work-Related Injuries and Illnesses and Recording Criteria Clarifications

Recordable injuries and illnesses (incidents that are work-related, a new case, and meet recording criteria) must be recorded on an OSHA 300 Form as seen in appendix B of this chapter. The OSHA 300 Form (Log of Work-Related Injuries and Illnesses) is the required document employers must use to log and track recordable incidents. On the 300 Form, incidents are logged by listing the employee's name, job title, date of injury or onset of illness, and location where the event occurred. Also, for each incident, employers must classify the incident as either a death, days away from work case, job transfer/restriction case, or other recordable cases. If the incident is a days-away case or job transfer/restriction case, then employers must log how many days were missed or were spent on job transfer/restriction. Lastly, on the 300 Form, the incident must be identified as an injury or an illness. If it is an illness, the incident must be classified as either a skin disorder, a respiratory condition, poisoning, hearing loss, or other illness. Employers must log recordable incidents on the 300 Log within seven calendar days of receiving information about the recordable incident.

When it comes to cases involving days away from work, OSHA provides some additional guidance on how to count days away. When logging the number of days away from work, employers must enter the number of calendar days away from work, not predicted workdays. This includes weekends and holidays. If the employee is out for an extended period of time, the employer must enter an estimate of the days the employee will be away, and update the day count when the actual number of days is known. The count of number of days away starts the day after the injury occurred or the illness began. If a physician recommends that the employee stay home from work and the employee comes to work anyway, the employer must record the incident as a days-away case and log the number of days away from work the physician recommended. If the physician does not recommend days away from work, but the employee stays home anyway or stays home longer than a physician recommends, the employer does not need to record it as a days-away case or only records the days away that a physician recommends. The limit to the number of days away that need to be recorded is 180. If employees miss more than 180 days due to an injury or illness, then only 180 days will be logged as days away on the 300 Form. Alternatively, the count of days away may end if the employee retires or leaves the company.

In addition to days-away clarifications, OSHA provides some clarification when adding restricted duty/job transfer cases to the 300 Form. Restricted duty means the employee is kept from performing one or more of his/her routine job functions or from working the full workday. Routine functions mean work activities the employee regularly performs at least once per week. If you assign employees to another job because of an injury or illness, the job transfer is treated in the same way as a job restriction. It is classified as the same type of case on the 300 Form. Job transfer/restriction days are counted and logged in the same way as days-away cases as seen in the previous paragraph. This guidance includes how to count cases, what days are counted, the 180-day cap, and so on.

Besides the general recordable criteria, there are a few specific criteria that make an incident a recordable injury or illness that must be added to the employer's 300 Log. All work-related needlestick injuries and cuts from sharp objects that are contaminated with another person's blood or other potentially infectious material are considered recordable injuries and must be added to the 300 Log. If an employee is medically removed under the medical surveillance requirements of an OSHA standard, you must record the case on the OSHA 300 Log. Some OSHA standards (thus employer programs) require employees to get regular medical check-ups to monitor the health effects of specific hazardous substances (e.g., lead). If an employee's medical results indicate he/she must be removed from exposure to a substance, then this medical removal meets recordable criteria, and the incident must be recorded on the 300 Form. If an employee's hearing test (audiogram) reveals that the employee has experienced a work-related Standard Threshold Shift (STS) (a.k.a. diagnosed hearing loss), in one or both ears, then the incident must be added to the 300 Log. If an employee is diagnosed with tuberculosis by a physician, from being exposed at work, it is a recordable case to be added to the 300 Log.

OSHA 300-A Form: Summary of Injuries and Illnesses and OSHA Incident Rates

In addition to the OSHA 300 Form, employers must complete the OSHA 300-A Form annually. The OSHA 300-A Form (appendix C) is the Summary of Work-Related Injuries and Illnesses. The OSHA 300-A Form summarizes the 300 Log for each calendar year. The 300-A Form annually documents the total number of deaths, total number of cases resulting in days away from work, total number of cases resulting in job transfer/restriction, and total number of other recordable cases. The 300-A Form also documents the total number of days away from work and total number of days of job restriction/ transfer that are a result of recordable incidents. The total number of injuries and types of illnesses are summarized on the 300-A Form as well. The 300-A Form must include company information such as location, industry description, number of employees, and the total number of hours worked by employees in the calendar year the form summarizes. The OSHA 300-A Form must be posted in the workplace annually. The certified (e.g., signed by a company executive) OSHA 300-A Form must be physically posted in the workplace from February 1 to April 30 every year. (summarizing last year's incidents).

Using the information in a 300-A Form, employers can calculate an OSHA incident rate for a given year. Equation 3.1 shows the OSHA total recordable incident rate (TRIR) equation.

Equation 3.1:

$$\text{TRIR} = \frac{\text{Number of recordable incidents} \times 200{,}000}{\text{Total hours worked}}$$

In the simplest form of the equation, the number of recordable incidents is the total number of all recordable incidents in the given calendar year. The number 200,000 is a constant in the equation. It represents 100 workers working 40 hours per week for 50 weeks in the year. Using this constant allows for the rate to be a more manager number (not so small) and for standard comparison when benchmarking against other companies. The total hours worked is the total number of hours worked by all employees in the given calendar year.

Another useful equation for comparing safety performance between companies is the days away/restricted time (DART) rate. The DART rate uses only the number of recordable cases that resulted in days away from work or restricted time/job transfer. The following equation shows the DART rate.

Equation 3.2:

$$\text{DART Rate} = \frac{\text{Number of cases resulting in days away or restrictions} \times 200{,}000}{\text{Total hours worked}}$$

This DART rate is often used as a measure of severity as compared to the TRIR. This is because the DART rate theoretically only considers injuries and illnesses that were relatively more severe and resulted in days off/

restricted time. Information on the OSHA 300-A Form can be used to calculate a TRIR and DART rate.

The TRIR and DART rate are useful for comparing the safety performance of companies. Using industry codes (e.g., SIC code, NAICS code), one can look up the average incident rates for a given industry. Then, companies can compare their rates to the industry average. In general, a good incident rate is below the industry average. Companies can also be compared against each other for safety performance by comparing their incident rates to each other.

OSHA 301 Form: Injury and Illness Incident Report

The OSHA 301 Form (appendix D) is called the Injury and Illness Incident Report. It is used for documenting and reporting injuries and illnesses when they happen in the workplace. The 301 Form allows for collecting information about the employee involved, information about the medical care needed for the injury or illness, and information about the case overall. The form allows for information to be documented about what happened, what the employee was doing, and what caused the injury or illness. The exact format of the OSHA 301 Form does not need to be used, but employers must have a formal established procedure for reporting and recording workplace injuries and illnesses. An incident report document is then used to complete the OSHA 300 Log with initial, necessary information.

Employees Who Must Be Included in Recordkeeping

Employers must include recordable injuries and illnesses on their OSHA 300 Log of all employees who are on payroll, whether they are labor, executive, hourly, salary, part-time, seasonal, or migrant workers. Employers must also record the recordable injuries and illnesses that occur to employees who are not on your payroll if the employer supervises these employees on a day-to-day basis. In no situation will an employee's recordable injury have to be logged on two different employers' 300 Forms (the employer who provides daily supervision and work direction should add it to their log).

Reporting Fatalities, Injuries, and Illnesses to the Government (OSHA)

Reportable incidents are those that must be reported directly to OSHA. The following work-related incidents are reportable to OSHA:

- Deaths
- In-patient hospitalization of one or more employees

- Loss of an eye
- An amputation

Deaths must be reported within eight hours of the incident. The other types of reportable incidents must be reported within twenty-four hours. Employers must report these incidents by one of the following methods:

- By telephone or in person to the nearest OSHA Area Office to the site of the incident
- By telephone to OSHA's call number at 1-800-321-6742
- By electronic submission on www.osha.gov

When reporting reportable incidents, Employers must provide the following information.

- The establishment, business, or company name
- The location of the incident
- The time of the incident
- The type of incident (death, hospitalization, loss of eye, amputation)
- The number of employees and names of those who were a victim of the reportable incident
- The name and phone number of the person reporting
- A brief description of the incident (what happened)

There are some specific things to clarify about reportable incidents. Employers do not have to report incidents related to motor vehicle crashes that occur on public roads unless it occurs in a construction work zone. Employers do not have to report incidents related to incidents on public transportation systems (e.g., airplane, train, subway, bus). An in-patient hospitalization is defined as a formal admission to the in-patient service of a hospital or clinic for care or treatment. So, visiting an emergency room does not necessarily mean the incident is reportable. It is possible employees receive observation or diagnostic testing in a hospital but are not admitted to the in-patient service of the hospital. For these instances, the incident would not be reportable. Employers should rely on the advice of a medical professional if unsure if an employee is admitted for in-patient service. Lastly, an amputation does not strictly mean the complete severing of a limb. Amputations include loss of a limb, loss of an external body part, fingertip amputations (with OR without bone loss), and medical amputations. Amputations do not include avulsions, enucleations, deglovings, scalpings, severed ears, or broken or chipped teeth. If employers are unsure if the incident classifies as an amputation, they should rely on medical diagnosis and recommendations from a physician or licensed health care professional.

Electronic Submission of Injury and Illness Records to OSHA

Most companies must submit OSHA incident recordkeeping forms to OSHA annually. Employers with 250 or more employees must electronically submit OSHA 300 Logs and OSHA 300-A Forms to OSHA on an annual basis. Employers with 20 or more employees but less than 250 employees who are in designated high-hazard industries must electronically submit OSHA 300 Logs and OSHA 300-A Forms to OSHA on an annual basis. Employers with less than twenty employees do not need to submit incident records to OSHA annually. Any employer must submit their recordkeeping form to OSHA upon request from OSHA. Those employers who must electronically submit records to OSHA must do so once a year, by the date of March 2. This submission will include forms of the previous calendar year (e.g., March 2, 2024 submission includes the year 2023 records). Recordkeeping forms must be submitted online at osha.gov/injuryreporting/.

REVIEW QUESTIONS

1. What is a recordable incident? What is a reportable incident?
2. What three things must be true for an incident to be a recordable incident?
3. What are five examples of recordable criteria?
4. Compare and contrast and provide examples of recordable incident general criteria and specific criteria.
5. Describe the purpose of a 300 Form and what information is required on it?
6. Describe the purpose of a 300-A Form and what information is required on it? When does it need to be posted? How do you calculate a TRIR? How do you calculate a DART rate? What are incident rates used for?
7. What incidents need to be reported to OSHA? What is the time limit on doing so?
8. What employers must electronically submit incident recordkeeping form to OSHA? When is the due date each year?

REFERENCES

Occupational Safety & Health Administration [OSHA]. (2017). Regulations (Standards-29 CFR 1904.0). Retrieved from https://www.osha.gov/laws-regs/regulations/standardnumber/1904/1904.0

Occupational Safety & Health Administration [OSHA]. (2020). Regulations (Standards-29 CFR 1904.1). Retrieved from https://www.osha.gov/laws-regs/regulations/standardnumber/1904/1904.1

Occupational Safety & Health Administration [OSHA]. (2014). Regulations (Standards-29 CFR 1904.2). Retrieved from https://www.osha.gov/laws-regs/regulations/standardnumber/1904/1904.2

Occupational Safety & Health Administration [OSHA]. (2017). Regulations (Standards-29 CFR 1904.4). Retrieved from https://www.osha.gov/laws-regs/regulations/standardnumber/1904/1904.4

Occupational Safety & Health Administration [OSHA]. (2001). Regulations (Standards-29 CFR 1904.5). Retrieved from https://www.osha.gov/laws-regs/regulations/standardnumber/1904/1904.5

Occupational Safety & Health Administration [OSHA]. (2001). Regulations (Standards-29 CFR 1904.6). Retrieved from https://www.osha.gov/laws-regs/regulations/standardnumber/1904/1904.6

Occupational Safety & Health Administration [OSHA]. (2001). Regulations (Standards-29 CFR 1904.7). Retrieved from https://www.osha.gov/laws-regs/regulations/standardnumber/1904/1904.7

Occupational Safety & Health Administration [OSHA]. (2001). Regulations (Standards-29 CFR 1904.8). Retrieved from https://www.osha.gov/laws-regs/regulations/standardnumber/1904/1904.8

Occupational Safety & Health Administration [OSHA]. (2001). Regulations (Standards-29 CFR 1904.9). Retrieved from https://www.osha.gov/laws-regs/regulations/standardnumber/1904/1904.9

Occupational Safety & Health Administration [OSHA]. (2019). Regulations (Standards-29 CFR 1904.10). Retrieved from https://www.osha.gov/laws-regs/regulations/standardnumber/1904/1904.10

Occupational Safety & Health Administration [OSHA]. (2001). Regulations (Standards-29 CFR 1904.11). Retrieved from https://www.osha.gov/laws-regs/regulations/standardnumber/1904/1904.11

Occupational Safety & Health Administration [OSHA]. (2017). Regulations (Standards-29 CFR 1904.29). Retrieved from https://www.osha.gov/laws-regs/regulations/standardnumber/1904/1904.29

Occupational Safety & Health Administration [OSHA]. (2001). Regulations (Standards-29 CFR 1904.31). Retrieved from https://www.osha.gov/laws-regs/regulations/standardnumber/1904/1904.31

Occupational Safety & Health Administration [OSHA]. (2020). Regulations (Standards-29 CFR 1904.33). Retrieved from https://www.osha.gov/laws-regs/regulations/standardnumber/1904/1904.32

Occupational Safety & Health Administration [OSHA]. (2014). Regulations (Standards-29 CFR 1904.39). Retrieved from https://www.osha.gov/laws-regs/regulations/standardnumber/1904/1904.39

Occupational Safety & Health Administration [OSHA]. (2019). Regulations (Standards-29 CFR 1904.41). Retrieved from https://www.osha.gov/laws-regs/regulations/standardnumber/1904/1904.41

APPENDIX A—EXAMPLES EXEMPT FOR RECORDKEEPING (OSHA 1904.5(B)(2))

1904.5(b)(2)	You are not required to record injuries and illnesses if . . .
(i)	At the time of the injury or illness, the employee was present in the work environment as a member of the general public rather than as an employee.
(ii)	The injury or illness involves signs or symptoms that surface at work but result solely from a non-work-related event or exposure that occurs outside the work environment.
(iii)	The injury or illness results solely from voluntary participation in a wellness program or in a medical, fitness, or recreational activity such as blood donation, physical examination, flu shot, exercise class, racquetball, or baseball.
(iv)	The injury or illness is solely the result of an employee eating, drinking, or preparing food or drink for personal consumption (whether bought on the employer's premises or brought in). For example, if the employee is injured by choking on a sandwich while in the employer's establishment, the case would not be considered work-related. **Note:** If the employee is made ill by ingesting food contaminated by workplace contaminants (such as lead), or gets food poisoning from food supplied by the employer, the case would be considered work-related.
(v)	The injury or illness is solely the result of an employee doing personal tasks (unrelated to their employment) at the establishment outside of the employee's assigned working hours.
(vi)	The injury or illness is solely the result of personal grooming, self medication for a non-work-related condition, or is intentionally self-inflicted.
(vii)	The injury or illness is caused by a motor vehicle accident and occurs on a company parking lot or company access road while the employee is commuting to or from work.
(viii)	The illness is the common cold or flu (Note: contagious diseases such as tuberculosis, brucellosis, hepatitis A, or plague are considered work-related if the employee is infected at work).
(ix)	The illness is a mental illness. Mental illness will not be considered work-related unless the employee voluntarily provides the employer with an opinion from a physician or other licensed health care professional with appropriate training and experience (psychiatrist, psychologist, psychiatric nurse practitioner, etc.) stating that the employee has a mental illness that is work-related.

APPENDIX B—OSHA 300 FORM

OSHA's Form 300 (Rev. 01/2004)

Log of Work-Related Injuries and Illnesses

Attention: This form contains information relating to employee health and must be used in a manner that protects the confidentiality of employees to the extent possible while the information is being used for occupational safety and health purposes.

Year 20___

U.S. Department of Labor
Occupational Safety and Health Administration

Form approved OMB no. 1218-0176

You must record information about every work-related death and about every work-related injury or illness that involves loss of consciousness, restricted work activity or job transfer, days away from work, or medical treatment beyond first aid. You must also record significant work-related injuries and illnesses that are diagnosed by a physician or licensed health care professional. You must also record work-related injuries and illnesses that meet any of the specific recording criteria listed in 29 CFR Part 1904.8 through 1904.12. Feel free to use two lines for a single case if you need to. You must complete an Injury and Illness Incident Report (OSHA Form 301) or equivalent form for each injury or illness recorded on this form. If you're not sure whether a case is recordable, call your local OSHA office for help.

Establishment name _____

City _____ State _____

Identify the person			Describe the case			Classify the case								

Public reporting burden for this collection of information is estimated to average 14 minutes per response, including time to review the instructions, search and gather the data needed, and complete and review the collection of information. Persons are not required to respond to the collection of information unless it displays a currently valid OMB control number. If you have any comments about these estimates or any other aspects of this data collection, contact: US Department of Labor, OSHA, Office of Statistical Analysis, Room N-3644, 200 Constitution Avenue, NW, Washington, DC 20210. Do not send the completed forms to this office.

Page ___ of ___

APPENDIX C—OSHA 300-A FORM

OSHA's Form 300A (Rev. 01/2004)

Summary of Work-Related Injuries and Illnesses

U.S. Department of Labor
Occupational Safety and Health Administration

Form approved OMB no. 1218-0176

Year 20____

All establishments covered by Part 1904 must complete this Summary page, even if no work-related injuries or illnesses occurred during the year. Remember to review the Log to verify that the entries are complete and accurate before completing this summary.

Using the Log, count the individual entries you made for each category. Then write the totals below, making sure you've added the entries from every page of the Log. If you had no cases, write "0."

Employees, former employees, and their representatives have the right to review the OSHA Form 300 in its entirety. They also have limited access to the OSHA Form 301 or its equivalent. See 29 CFR Part 1904.35, in OSHA's recordkeeping rule, for further details on the access provisions for these forms.

Number of Cases

Total number of deaths

(G)

Total number of cases with days away from work

(H)

Total number of cases with job transfer or restriction

(I)

Total number of other recordable cases

(J)

Number of Days

Total number of days away from work

(K)

Total number of days of job transfer or restriction

(L)

Injury and Illness Types

Total number of . . .
(M)

(1) Injuries

(2) Skin disorders

(3) Respiratory conditions

(4) Poisonings

(5) Hearing loss

(6) All other illnesses

Establishment Information

Your establishment name _____

Street _____

City _____ State ____ ZIP ____

Industry description (e.g., Manufacture of motor truck trailers) _____

Standard Industrial Classification (SIC), if known (e.g., 3715) _____

OR

North American Industrial Classification (NAICS), if known (e.g., 336212) _____

Employment Information (If you don't have these figures, see the Worksheet on the back of this page to estimate.)

Annual average number of employees _____

Total hours worked by all employees last year _____

Sign here

Knowingly falsifying this document may result in a fine.

I certify that I have examined this document and that to the best of my knowledge the entries are true, accurate, and complete.

_____
Company executive   Title

_____
Phone   Date

Post this Summary page from February 1 to April 30 of the year following the year covered by the form.

Public reporting burden for this collection of information is estimated to average 58 minutes per response, including time to review the instructions, search and gather the data needed, and complete and review the collection of information. Persons are not required to respond to the collection of information unless it displays a currently valid OMB control number. If you have any comments about these estimates or any other aspects of this data collection, contact: US Department of Labor, OSHA Office of Statistical Analysis, Room N-3644, 200 Constitution Avenue, NW, Washington, DC 20210. Do not send the completed forms to this office.

APPENDIX D—OSHA 301 FORM

OSHA's Form 301
Injury and Illness Incident Report

Attention: This form contains information relating to employee health and must be used in a manner that protects the confidentiality of employees to the extent possible while the information is being used for occupational safety and health purposes.

U.S. Department of Labor
Occupational Safety and Health Administration

Form approved OMB no. 1218-0176

This *Injury and Illness Incident Report* is one of the first forms you must fill out when a recordable work-related injury or illness has occurred. Together with the *Log of Work-Related Injuries and Illnesses* and the accompanying *Summary*, these forms help the employer and OSHA develop a picture of the extent and severity of work-related incidents.

Within 7 calendar days after you receive information that a recordable work-related injury or illness has occurred, you must fill out this form or an equivalent form. Some state workers' compensation, insurance, or other reports may be acceptable substitutes. To be considered an equivalent form, any substitute must contain all the information asked for on this form.

According to Public Law 91-596 and 29 CFR 1904, OSHA's recordkeeping rule, you must keep this form on file for 5 years following the year to which it pertains.

If you need additional copies of this form, you may photocopy and use as many as you need.

Completed by _____

Title _____

Phone (____) ____ - ____   Date ____ / ____ / ____

Information about the employee

1) Full name _____

2) Street _____

City _____ State _____ ZIP _____

3) Date of birth ____ / ____ / ____

4) Date hired ____ / ____ / ____

5) ☐ Male ☐ Female

Information about the physician or other health care professional

6) Name of physician or other health care professional _____

7) If treatment was given away from the worksite, where was it given?

Facility _____

Street _____

City _____ State _____ ZIP _____

8) Was employee treated in an emergency room?
☐ Yes ☐ No

9) Was employee hospitalized overnight as an in-patient?
☐ Yes ☐ No

Information about the case

10) Case number from the Log _____ (Transfer the case number from the Log after you record the case.)

11) Date of injury or illness ____ / ____ / ____

12) Time employee began work _____ AM / PM

13) Time of event _____ AM / PM ☐ Check if time cannot be determined

14) **What was the employee doing just before the incident occurred?** Describe the activity, as well as the tools, equipment, or material the employee was using. Be specific. *Examples:* "climbing a ladder while carrying roofing materials"; "spraying chlorine from hand sprayer"; "daily computer key-entry."

15) **What happened?** Tell us how the injury occurred. *Examples:* "Worker was spraying chlorine with chlorine when gasket broke during replacement"; "Worker developed soreness in wrist over time"; "When ladder slipped on wet floor, worker fell 20 feet."

16) **What was the injury or illness?** Tell us the part of the body that was affected and how it was affected; be more specific than "hurt," "pain," or sore." *Examples:* "strained back"; "chemical burn, hand"; "carpal tunnel syndrome."

17) **What object or substance directly harmed the employee?** *Examples:* "concrete floor"; "chlorine"; "radial arm saw." *If this question does not apply to the incident, leave it blank.*

18) **If the employee died, when did death occur?** Date of death ____ / ____ / ____

Chapter 4

Walking-Working Surfaces

Chapter 4 was written by Travis Spagnolo.

INTRODUCTION AND SCOPE

Each year, one of the leading causes of death and serious injury in the workplace is falls from heights. In January of 2017, OSHA issued the Final Rule for the new walking-working surfaces regulation, 1910 Subpart D. This new regulation was developed to provide guidance for employers to prevent falls from heights and on the same working level. Prior to the inception of this regulation, very minimal regulatory guidance existed that helped to prevent falls in industrial settings such as manufacturing facilities, warehouses, and commercial properties. This regulation affects a wide range of personnel who may be tasked with performing any number of duties at heights such as maintenance personnel, equipment operators, other skilled trades, and property managers.

Language in this regulation is intended to align closely with OSHA's fall protection in construction regulation, 1926 Subpart M. However, one key difference in 1910 Subpart D compared to 1926 Subpart M is the height at which fall protection is required. The construction regulations call for fall protection to be provided at a height that is 6 feet above a lower level, while 1910 Subpart D requires fall protection to be used when an employee is only 4 feet above a lower level. The walking-working surfaces regulation also identifies training requirements which are similar to those listed for employers covered under 1926 Subpart M performing construction activities.

Beyond the 4-foot rule in general industry, employers must comply with specific fall protection requirements. Guardrails must meet specific design and dimension standards. Personal fall arrest systems (PPE used to protect employee in the event of a fall) must be properly rated and arranged in a way to afford protection. Alternate fall protection methods must meet minimum requirements set by OSHA. Stairways and ladders must meet specific criteria set by OSHA, and employees must receive training to avoid fall hazards when potentially exposed to them in the workplace.

Nearly all facilities have an area where their employees will experience exposure to a potential fall. It is up to the employer to identify these tasks and potential exposures that exist in the workplace. Given the implementation of OSHA's recent walking-working surfaces regulation, it should now be easier than ever to formulate a plan that significantly reduces fall hazards in the workplace.

The OSHA standards covered in this chapter include:

- 29 CFR 1910 Subpart D:
  - 29 CFR 1910.22—General requirements
  - 29 CFR 1910.23—Ladders
  - 29 CFR 1910.25—Stairways
  - 29 CFR 1910.28—Duty to have fall protection and falling object protection
  - 29 CFR 1910.29—Fall protection systems and falling object protection criteria and practices
  - 29 CFR 1910.30—Training requirements
  - 29 CFR 1910.140—Personal fall protection systems
  Not included in this chapter:

- Under 29 CFR 1910 Subpart D:
  - Requirements of safety nets
  - 29 CFR 1910.21—Scope and definitions
  - 29 CFR 1910.24—Step bolts and manhole steps
  - 29 CFR 1910.26—Dockboards
  - 29 CFR 1910.27—Scaffolds and rope descent systems

NARRATIVE

Unlike many of his friends in school, Bill is undecided about what to do after graduating high school. The thought of going to a four-year college for something that he is not passionate about did not interest Bill. While his friends are preparing to go to school to learn subjects like accounting and chemistry, Bill is busy figuring out what he wanted to do with his life. There was a time when Bill felt left out that he was opting to not go to college somewhere, but he also knows deep down that he wanted to work with his hands for a living; that is where his true passion existed. After all, his dad owned an automotive repair shop, and Bill always enjoys helping out during his free time.

Bill enjoys fixing things. He likes taking things apart and finding what is wrong. It is part of why he enjoys working in his dad's repair shop, and it was that skill that helped him determine the next step in his life. One day, near the end of his senior year of high school, Bill is working with his dad in the garage of their repair shop. They notice that the air conditioning unit in the building is no longer working. Bill, excited about the opportunity to prove to his dad that he can fix the air conditioner, jumped right in and started taking it apart. After about an hour or so of troubleshooting and disassembly, Bill identifies a number of items on the unit in need of repair. "Our filters are clogged, Dad, and the evaporator coil isn't working right either," Bill told his dad. After spending some more time fixing the unit and getting it back up and running, Bill's father gives him some great advice. "You know son, you should consider going to trade school. There's always a market for HVAC technicians."

At that moment, it dawned on Bill that he knew what he wanted to do after high school. That evening, Bill applies at a local trade school to enter a two-year HVAC program. He would start in the fall.

Bill progresses through the HVAC program with relative ease. His time spent working on cars really helped him to be prepared for such a hands-on curriculum. A big part of the program is spent working with a trained HVAC technician in the field. Bill is occasionally asked to do some work from a ladder or a scissor lift, but he is rarely ever asked to do work at heights. If he needs to work on a roof, he simply is told to stay away from the roof's edge. Bill typically feels pretty safe when he needs to work from a roof or on a scissor lift.

Bill is now at the end of the HVAC program; instructors feel confident that Bill can begin working on his own doing some small repairs. The school has a number of customers with whom they have agreements with to allow students to work on their HVAC equipment if something breaks down. As this individual work becomes part of Bill's normal day, he finds himself working on roofs by himself more frequently. Although he likes the ability to work independently, his parents are concerned that some of this work may be unsafe. "Don't you wear a harness when you go up there, Bill?" his dad asks one night at dinner. "I stay away from the edge Dad, it's fine. Besides, they have never even shown us how to wear those things. I feel better just being aware of my surroundings," Bill responded.

With just six weeks left in the program, Bill is looking forward to graduating and starting to make some money in the workforce. Bill has accepted a job with an HVAC company that services customers and retail businesses in his area. To share his good news and to celebrate, Bill's friends from high school are all planning on coming home from their respective colleges for a weekend. His friend Mike arranges for them all to go away to his family's cabin where they would enjoy some outdoor activities while catching up with each other.

Bill has a full day of repairs his instructor lined up for him to do on Friday before he headed away for the weekend. He tells his friends his car is packed, and he would make his way to the cabin immediately after finishing his last repair. This Friday is a particularly cold day in the middle of March. Temperature is near freezing, which tends to make working on roofs a little slippery, especially with recent precipitation.

Bill arrives at his first stop earlier than normal. There are multiple rooftop units in need of service, and Bill had never been to this facility before, so he feels it is best to start early. After accessing the roof to perform his repairs, Bill feels a little uneasy. Most of the condensing units he was servicing are about 5 or 6 feet from the edge of the roof or closer, and there are patches of ice that were difficult to identify on the white roof surface. Bill initially thinks about calling his instructor to ask what he should do, but then he thinks about trying to finish work early that day to get to the cabin quicker. Bill decides to just be extra careful and proceed with performing the repairs.

The first unit Bill services gives him some trouble. He is trying to replace a belt for the fan motor, but he is running into problems. It didn't help; he was also nervous about being so close to an unprotected roof edge. As Bill stood up to retrieve a different tool from his tool bag, his left foot steps on a patch of ice that he did not see. Bill loses his footing and was unable to regain his balance. He lands on his side, rolls once, and falls 35 feet off the roof to the ground below.

Bill lands on the sidewalk in the back of the facility. It takes about an hour for somebody to finally see him lying in a puddle of blood and unresponsive. An employee for the company immediately calls 911 after noticing Bill, but by the time the paramedics arrive, it is already too late. Bill is pronounced dead at the scene. Unknown at the time, Bill's friends are already embarking on their weekend adventure to celebrate his new job and accomplishments. Unfortunately, Bill never makes it to celebrate with them.

SUMMARY OF OSHA STANDARDS

What Is Covered?

Understanding the requirement for fall protection first involves an employer recognizing what OSHA classifies as a walking working surface and what is covered under 1910 Subpart D. A walking working surface is "any horizontal or vertical surface on or through which an employee walks, works, or gains access to a work area or workplace location." This can be any location in a facility, and although the regulation is primarily designed to provide regulatory guidance for work that is performed on elevated walking-working surfaces, Subpart D 1910 can be applied much more broadly. Aisles, passageways, platforms, workrooms, and any other walkways are also covered in the regulation.

OSHA requires that the employer perform regular inspections of walking-working surfaces to determine that they are safe, kept clear, and free of hazards. A very basic example of a common walking working surface that will be present in most facilities is the sidewalk outside the building. It is the responsibility of the employer to ensure that their sidewalk is kept clear, especially if employees are using it regularly. For example, when snowy and icy conditions are likely to occur in the winter months, part of the employer's walking-working surfaces program must include a regular inspection procedure to evaluate the sidewalk for ice and treat it as necessary. Consider the same application inside of a manufacturing facility. Any location where employees may be expected to be either walking or working qualifies as a walking working surface. For example, food processing plants have high-pressure water jets and water hoses to clean surfaces and products continuously. This process may create the hazard of both overspray and product, which can become slippery when wet, landing in highly traveled walking paths. Part of the employer's responsibility is to ensure that they are inspecting and clearing these walking paths regularly to eliminate slip, trip, and fall hazards. In facilities where there is limited ability to control material or parts of the manufacturing process from landing in walking paths, a potential control measure is installing non-slip walking mats.

Part of OSHA's enforcement of 1910.22(d)(2) calls for the employer to block or prevent access to areas where walking-working surfaces are not safe for employee use. So technically, those icy sidewalks mentioned previously should be blocked until they can be cleared. OSHA's application of 1910 Subpart D is broad and covers most areas where employees will be walking and working in a facility, as the name of the regulation suggests. Ensuring clear access to any of these areas is also critical. For example, doorways and walking paths throughout the facility must not be blocked with debris or material storage. Many facilities will establish painted lines on the floor to mark walkways and appropriate storage areas. Employers must train their employees to only store material in approved storage locations so as to not block walking paths and aisleways. The best practice is to ensure that aisleways and walking paths are maintained with 28 inches of clear space, which is the required width that must be maintained in exit routes.

Many facilities have elevated mezzanines or platforms where material and equipment may be kept for storage purposes. These surfaces are considered walking-working surfaces and are covered under 1910.22(b) to ensure the safe loading of walking-working surfaces. The creation of these mezzanines and loading platforms often occurs when expansions or additions to the original facility occur. When construction is finished, or when the building becomes occupied, it is imperative that load ratings be established to determine if loads applied to the floor do not pose a potential hazard that could lead to the floor, mezzanine, or platform collapsing or falling. In situations where the structural integrity or load ratings of walking-working surfaces is unknown, the best course of action is to utilize the services of a qualified person such as a structural engineer to determine the capacity of the surface. OSHA does not list the method that is required to determine the capacity of a surface which is loaded; however, they do require that the employer take the necessary steps to obtain the needed information.

Employers should not only focus on walking paths and loading platforms. This regulation also addresses any work that is to be performed at heights. OSHA requires in general industry that employers provide fall protection at heights more than 4 feet above a lower level. This applies to roofs, unprotected decks and platforms, or any other elevated surfaces such as the top of the equipment. Once an employer identifies a task that is to be performed at heights, installing fall protection is the next critical step to protect employees.

Fall Protection Solutions

There are two primary categories of fall protection which can be selected to provide fall protection for employees. The first, conventional fall protection, will either stop an employee from striking the lower level or eliminate a fall altogether. These are generally considered to be the most effective methods of control for fall protection. Conventional methods include guardrail systems, personal fall arrest systems (PFAS), and safety nets.

The second category of fall protection solutions is referred to as alternative methods. These methods do not eliminate or control a fall. Instead, they rely on a combination of training, supervision, and visual markings (administrative controls) to ensure employees do not go near the unprotected edge. Some alternative systems are only permitted to be used on a low-sloped roof in limited applications.

After determining that employees are exposed to a fall greater than 4 feet above a lower level, the employer must select which method of fall protection to provide for employees. In utilizing the commonly referenced hierarchy of controls in safety, employers should first attempt to eliminate a hazard. The only method of fall protection that would eliminate a fall hazard involves the installation of a guardrail system (or similar adequate barrier). Proper installation will enclose an unprotected edge, thus preventing an employee from falling.

Guardrail systems consist of a top rail, mid-rail, supporting post or stanchion, and falling object protection if necessary. Figure 4.1 shows a standard-compliant guardrail system. Top rails must be installed at a height of 42″ +/− 3″ above the walking working surface. Top rails must be engineered to support a force of 200 pounds applied to the railing from any direction. Mid-rails are intended to prevent accidental falls underneath the top rail and through the guardrail system. Mid-rails must be installed halfway between the height of the top rail and walking working surface. Mid-rails within a guardrail system must be engineered to support 150 pounds of force applied to the railing from any direction.

The supporting or bracing forces for guardrail systems come from the strength that is generated from the structural support posts or stanchions. Although most industrial facilities will have metal or steel guardrail systems installed, systems that are installed temporarily and are constructed of wood will require careful installation given the spacing of the support posts or stanchions. A wooden guardrail system is seen in figure 4.2. When spacing between stanchions reaches more than 8 feet apart, guardrails systems tend to become weaker and may not support the required forces (200 pounds for top rail and 150 pounds for mid-rail). Best practice is to maintain spacing of no more than 6–8 feet between support stanchions to ensure that guardrails are braced properly.

Figure 4.1   Metal/Steel Guardrail System.

Figure 4.2   Wooden Guardrail System.

Any railing that is part of the guardrail system must be kept free of cut and laceration hazards. Examples of these hazards commonly identified on guardrails include nails and sharp metal edges. To assist in ensuring that guardrails do not create these cut hazards, materials such as steel banding and plastic banding are prohibited from being used as a top or mid-rail. Additionally, top and mid-rails must be a minimum of ¼ inch in thickness or diameter.

When guardrail systems are used in areas where there is potential for material to fall or roll off the edge of the walking-working surface onto employees below, falling object protection must be installed. Toe boards are common methods to prevent material from falling. These devices are installed onto guardrail systems and must be a minimum of 3½ inches in height, measured from the walking working surface. No more than ¼ inch of gap between the walking working surface and bottom of the toe board is permitted. Toe boards must support a minimum of fifty pounds of force applied in any direction. Other options to provide overhead protection include the installation of debris nets or screens onto the guardrail system. Barricading the area on the lower level using methods such as danger tape or other means to physically block access to an area where overhead hazards are present is also acceptable. If utilizing ground barricades, it is critical to ensure that enough space of a buffer zone is given to account for how far away from the edge of the upper level that material might fall outward and onto the ground below.

If it is not feasible to install guardrail systems, employers should consider another method of conventional fall protection: PFAS. Figure 4.3 shows a PFAS. A PFAS includes the use of a full-body harness, a connecting device, and an anchorage device. This method of fall protection does not eliminate falls. Instead, it controls falls and prevents the employee involved in the fall event from striking the lower level. For PFAS to work effectively, employees must be properly trained in the correct use of equipment, equipment must be in serviceable condition, and anchorage devices and surfaces must be capable of supporting the employee involved in the fall.

When installing a PFAS, a qualified person must determine a suitable surface to affix an anchorage device. Examples of commonly used surfaces where anchorage devices are affixed include structural steel beams, concrete decks, and roof surfaces that range from wood to metal or concrete. The anchorage device that is installed to any of these surfaces must also be designed to support 5,000 pounds of force. Numerous types and styles of anchors are commercially available for purchase depending on the application of use and surface to which you are anchoring.

Figure 4.3   Employee Attached to an Overhead Anchorage Point.

It is critical to determine that the surface to which your anchor is affixed is capable of being used as part of a fall arrest system. Simply using an anchorage device capable of supporting 5,000 pounds of force does not mean that the anchor will be effective in a fall event. For example, an engineered anchor point wrapped around a weak surface such as the leg of a table will not be effective. Despite the anchorage device being capable of supporting the necessary force in a fall, the surface to which it was affixed was not. Qualified persons such as structural engineers should be involved in the process of determining which surfaces are suitable to install an anchor.

After installing an anchorage device onto an appropriate surface, the employee must connect their harness to the anchor using a connecting device, which is equipped with either double-locking snap hooks or double-locking carabiners. There are multiple styles of connecting devices. Common connecting devices include a shock-absorbing lanyard, typically 6 feet in length, and a self-retracting lifeline. Other devices such as a rope grab may be used as part of a vertical lifeline system. Some connecting devices such as travel restraints are not considered part of fall arrest because they prevent an employee from reaching an edge or they limit from where an employee could fall, or they prevent free fall altogether. This is also termed fall restraint. Travel restraints are also used as part of positioning systems while climbing fixed ladders.

The combination of equipment that is used in a PFAS is intended to limit the maximum forces generated on the employee to 1,800 pounds, while limiting freefall distance to a maximum length of 6 feet. This is why the maximum length of shock-absorbing lanyard that is available on the market is 6 feet. A freefall greater than 6 feet would generate forces on the employee falling to more than 1,800 pounds.

Connecting devices such as shock-absorbing lanyards, as seen in figure 4.4, have deceleration devices built into the equipment which help to decelerate or cushion the employee's fall by having a breakaway feature built into the lanyard. After the end of the freefall, employees do not undergo an immediate and sudden stop. This would cause fall forces to exceed 1,800 pounds. Instead, after the freefall occurs, the employee's deceleration or breakaway device releases up to a maximum of 3½ feet to decelerate the employee's momentum and safely stop the fall.

One disadvantage to using shock-absorbing style connecting devices is the increased fall clearance distances. Consider that an employee will freefall 6 feet and will also have 3½ feet of deceleration device that will break away in a fall. Add to that length the average height of a worker (~6 feet) and a safety factor of about 3 feet given potential stretch of the harness and other considerations. When you combine those measurements, an employee will need a fall clearance distance of 18½ feet from the point of anchorage for a commonly used 6-feet shock-absorbing lanyard to be effective. See figure 4.5. Not only does OSHA

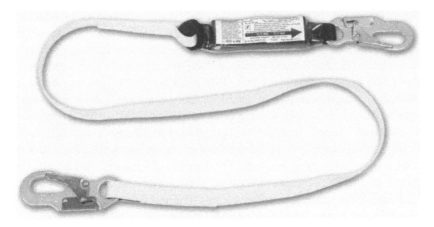

Figure 4.4   Shock-Absorbing Lanyard.

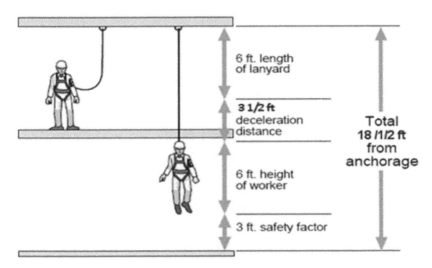

6 ft. length
of lanyard

3 1/2 ft
deceleration
distance

6 ft. height
of worker

3 ft. safety factor

Total
18 /1/2 ft
from
anchorage

Figure 4.5   Fall Clearance Distance Calculation.

require that employers provide their employees with fall protection when work-ing at heights of 4 feet or greater, but they also require that the method of fall protection that is provided prevents them from striking a lower level.

Consider an employee working on top of a piece of equipment that is 6 feet above the ground. He attaches himself to an anchorage device that is another 4 feet above his head using a 6-feet shock-absorbing lanyard. In this example, the employee is anchored 10 feet above the ground, and his connecting device requires a clearance distance of 18½ feet to stop his fall safely. This employee would still strike the ground in a fall in this scenario.

To account for fall clearance distances, many employers provide self-retracting lifelines to their employees to use when working at heights. Figure 4.6 depicts a self-retracting lifeline. These devices may or may not be equipped with deceleration devices. Self-retracting lifelines function in a manner that is similar to a seatbelt. When an amount of force is applied to a seatbelt at a particular rate of speed, it will immediately lock and stop the passenger. The same application applies to personal fall arrest when using a self-retracting lifeline. Because the stop happens almost instantaneously to the fall, the forces applied on the employee's body are significantly less than that of a shock-absorbing lanyard. The freefall distance is also significantly reduced. In the example we discussed above with the employee anchored 10 feet above the ground, a self-retracting lifeline would be effective in preventing the employee from striking the ground.

The third type of conventional fall protection is safety nets. These are not commonly used in manufacturing or industrial facilities; however, they are an option for employers to keep their employees safe. Safety net requirements are omitted from this chapter.

Figure 4.6   Self-Retracting Lifeline.

Alternative Methods

Under 1910 Subpart D, there are two key alternative methods of fall protection to consider besides the conventional methods. The first is the use of a designated area. The second alternative method is the employer's implementation of a work rule for temporary and infrequent tasks. Remember, alternative methods of fall protection are not preventing or controlling falls. They are simply systems that can be utilized to provide warning of fall hazards to employees. The success of using alternative methods of fall protection relies heavily on the employer having an effective fall protection training program. These are administrative controls for reducing risk and should only be considered if conventional fall protection is not feasible.

The use of these alternative methods of fall protection is only applicable on low-sloped roofs when temporary and infrequent work is occurring. Low-sloped roofs are roofs with a slope or pitch that is 4:12 or less. For example, a flat roof qualifies as a low-sloped roof.

Questions often arise from employers regarding what OSHA considers "temporary and infrequent" work tasks. Although 1910 Subpart D does not specifically define what constitutes a task that is performed temporarily or infrequently, industry practice considers that short tasks taking around 1–2 hours are temporary work tasks. A task such as replacing a belt for a fan motor on a condensing unit is typically a short task that would qualify as being temporary. Infrequent tasks are those performed sporadically or on occasion when needed. For example, performing an annual preventive maintenance inspection on a piece of rooftop equipment is an infrequent task. Performing a repair that is needed on occasion, such as the example of replacing a belt for a fan, is also considered infrequent.

To be exempt from using conventional fall protection when working on a low-sloped roof, these tasks must be both temporary *and* infrequent. Industrial facilities with ammonia refrigeration systems typically send maintenance employees onto the roof of the facility daily to check for any concerns with the ammonia lines and piping. Although this task may be temporary and take just a few minutes, it would not be considered infrequent because employees would be accessing the roof daily. In this instance, these employees must be provided with conventional fall protection by their employer.

Designated areas, commonly referred to as warning line systems when used in construction, create a safe working zone on an unprotected roof where employees are not required to wear personal fall arrest equipment. Figure 4.7 shows a designated area. Using this system, the employer is not required to install other methods of conventional fall protection like guardrails or safety nets either. This alternative method of fall protection is intended for tasks that

Figure 4.7   Designated Area on Roof.

are performed on low-sloped roofs that are temporary and infrequent and are
performed between 6 and 15 feet from the roof edge.

Using devices such as cones to hold up and support a warning line, the
designated area must fully enclose the employees working on the roof. The
warning line that is part of the designated area is permitted to be made out
of rope, wire, or chain and must have a minimum breaking strength of 200
pounds. Installation of the warning must not be closer than 6 feet from the
edge of the roof and the warning line must maintain the height of 34–39
inches above the walking working surface.

In designated areas, a common mistake during installation is the failure to
account for ladder access. If accessing the roof from a fixed ladder or extension ladder on the side of the roof, employees are not permitted to walk freely
into the designated area between the unprotected edge and the designed area.
An enclosed walkway must be installed using the warning line that connects
to the designated area. Figure 4.8 depicts a designated area walkway. This
provides employees a path to and from the protected and enclosed designated area during roof access and egress. If employees go beyond or outside
the designated area at any point during access, egress, or while performing
temporary and infrequent work, the employee must be protected using conventional fall protection.

If temporary and infrequent work is performed on a low-sloped roof
15 feet or more away from the unprotected roof edge, employers are not
required to use any fall protection and are not required to install a designated

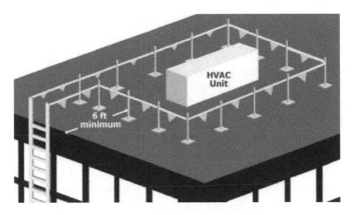

Figure 4.8   Protected Walkway Leading into a Designated Area.

area. Instead, employers have the option to implement and enforce a work rule for tasks that are both temporary and infrequent.

OSHA does not give guidance on examples of work rules, but they do stipulate that the work rule must prevent employees from traveling within 15 feet from the unprotected roof edge. Consider that when utilizing this method, there is no requirement for fall protection or a designated area. But, if employees travel within 15 feet from the roof edge, they are exposing themselves to a fall hazard and the employer is violating OSHA's walking-working surfaces regulation.

Many employers have developed work rules by making alterations to their existing roofs. A common example that many business owners choose to implement is a painted line on the surface of the roof that is at least 15 feet from the roof edge. Employees are then trained that as long as they are performing tasks that are temporary and infrequent, and remain inside of the painted line, no fall protection is required. Training should include discussions on options for conventional fall protection if the employee is required to go beyond the marked line on the roof surface. This approach works well for businesses where rooftop equipment is not located near roof edges.

Other employers have installed walking mats on roofs that lead to any potential equipment that an employee may need to service. This approach provides a clear pathway for employees to follow after accessing the roof to lead them straight to their work area while also remaining more than 15 feet from the roof edge.

Because OSHA does not list any guidance for establishing work rules for temporary and infrequent work, there are a number of creative ways that employers can develop a plan for their employees when performing these tasks. Work rules for temporary and infrequent work tend to fail when employees do not receive adequate training and when the rule is not enforced.

It should also be noted that employees *are* required to develop and implement a work rule of some variety. Simply communicating to employees to not walk within 15 feet of the roof edge is not an adequate or acceptable work rule. The employer must demonstrate how they have trained employees to recognize at what point on the roof they are nearing or have stepped within 15 feet from the roof edge.

Stairways

Under 1910 Subpart D, OSHA provides regulatory guidance for standard stairs, spiral stairs, ship stairs, and alternating tread-type stairs. Two critical components of stairs include the treads and risers. When evaluating stairs in a facility, many of OSHA's requirements for stairs are based on the number of treads and risers, as well as their size. When determining if stairs are safe and compliant with the regulations, evaluating the risers and treads is the ideal starting point.

A riser "is the upright (vertical) or inclined member of a stair that is located at the back of a stair tread or platform and connects close to the front edge of the next higher tread, platform, or landing." A tread is "a horizontal member of a stair or stairway, but does not include landings or platforms." Consider walking up a normal flight of stairs in your home. What your feet step on as you are climbing the stairs is considered the tread. The vertical piece at the back of the tread that connects to the next tread, or step, is the riser. Figure 4.9 illustrates stair treads and risers.

Regardless of the style of stairs that are installed at the facility, the height of all risers and the depth of all treads on the staircase must be uniform for each set of stairs. Tables 4.1, 4.2, 4.3, and 4.4 provide the acceptable uniformed

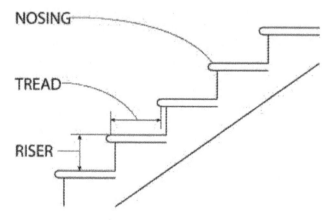

Figure 4.9   Diagram of Stair Treads and Risers from 1910.25(c)(4).

Table 4.1

Standard Stairs	
Installation Angle	30–50 degrees from horizontal
Maximum Riser Height	9½ inches
Minimum Tread Depth	9½ inches
Minimum Width of Stair Tread from Vertical Members (walls, stair rail systems, etc.)	22 inches

sizing for treads and risers for each type of stairs that may be installed in a facility. These tables also list other key dimensions as necessary for the style of the staircase such as the required clearance for headroom when walking on stairs or onto platforms from stairs. Figure 4.10 shows a standard stairway, figure 4.11 shows a spiral stairway, and figure 4.12 shows ship-style stairs.

The minimum clear width referenced in the table above refers to the width of the stair tread, as well as spacing between handrails or stair rail systems which are installed on either side of the tread.

As can be identified in Table 4.1 and Figures 4.10 and 4.12, there are differences between ship stairs and standard stairs. Risers are open on ship stairs, while they are closed on standard stairs. Additionally, based on the angle of installation, ship stairs are much steeper and can also be smaller and

Figure 4.10   Standard Stairs.

Table 4.2

	Spiral Stairs
Maximum Riser Height	9½ inches
Minimum Tread Depth	7½ inches measured at a point 12 inches from the narrower edge
Minimum Clear Width	26 inches
Minimum Headroom above Spiral Stair Treads	6 feet, 6 inches

narrower than standard stairs. Many facilities use ship stairs to access platforms and equipment. Commercial properties such as offices and high-rise residential buildings such as multistory apartments install ship stairs during construction of the building to be used as roof access.

It is recognized by OSHA that the type of stairs considered safest for employee travel between walking-working surfaces are the standard stairs. In cases where regular and routine travel is anticipated between walking-working surfaces using stairs, employers must, if feasible, install standard stairs. The installation and use of ship stairs, spiral stairs, and alternating

Figure 4.11   Spiral Stairs.

Table 4.3

Ship Stairs	
Installation Angle	*50–70 degrees from horizontal*
Acceptable Range of Riser Heights	6½ to 12 inches
	Risers on ship stairs are open (as shown in picture below)
Minimum Tread Depth	4 inches
Minimum Width of Stair Tread	18 inches

tread-type stairs (figure 4.13) are only permitted if the employer can prove that it is infeasible to install standard stairs at the location needed for access to another level or walking working surface. As a general rule, consider table 4.5 to remember what types of stairs are permitted at different angles installed from the horizontal ground surface.

Accessing and climbing any type of stairs also creates numerous safety concerns to consider. Some considerations that apply to all stairs include requirements for handrails, intermediate members, stair rail systems, and safe and clear access onto platforms.

After stepping off the final step and onto a working platform, the platform and landing area must have fall protection provided if more than 4 feet above a lower level. Protective solutions such as guardrails, safety nets, or personal fall arrest must be installed in accordance with what we have already

Figure 4.12   Ship Stairs.

Table 4.4

	Alternating Tread-Type Stairs
Installation Angle	Treads installed in series at a slope of 50–70 degrees from the horizontal
Maximum Riser Heights	9½ inches
	If riser heights are less than 9½ inches, the risers are permitted to be kept open, similar to ship stairs
Minimum Tread Depth	8½ inches
Minimum Width of Stair Tread	7 inches
Acceptable Distance between Handrails	17–24 inches

discussed in this chapter. Typically, the most commonly used form of fall protection on stair landings and platforms is a guardrail system.

Access onto these platforms must be clear and the platform must be appropriately sized. Platforms and landings must be as wide as the last stair and at least 30 inches in depth as measured in the direction of travel. When gates or doors are installed onto stairs and lead to a platform, the installation of the door or gate must not open in a manner that reduces the depth of the platform to be less than 20 or 22 inches depending on the date that the stairs were installed. Stairs installed before January 17, 2017, are permitted to have doors or gates reduce the platform depth to a maximum of 20 inches. Any stairs installed on or after January 17, 2017, must not allow for doors or gates to reduce the platform depth to any less than 22 inches.

Figure 4.13   Alternating Tread-Type Stairs.

Table 4.5

Access Depending on Angle from Horizontal	
Less than 30 degrees	Install a ramp, not stairs
30–50 degrees	Standard stairs
50–70 degrees	Ship stairs or alternating tread-type stairs
60–90 degrees	Ladders (fixed or straight/extension)

Headroom and vertical clearance from any overhead obstruction while on any stair tread must be a minimum of 6 feet, 8 inches. This applies to all stair types except spiral stairs, which require overhead clearance of 6 feet, 6 inches.

All stairways that are equipped with at least three treads and four risers must have handrails and stair rail systems installed in accordance with Table D-2 of 1910.28. This table provides regulatory guidance for how handrails are to be installed based on the width of the stair treads and the number of open sides that are installed. Employers should be aware that regardless of

Table 4.6

Stairway Handrail Requirements (1910.28(b)(11)(ii))				
Stair Width	Enclosed	One Open Side	Two Open Sides	With Earth Built Up on Both Sides
Less than 44 inches	At least one handrail	One stair rail system with handrail on open side	One stair rail system each open side	
44 inches to 88 inches	One handrail on each enclosed side	One stair rail system with handrail on open side and one handrail on enclosed side	One stair rail system with handrail on each open side	
Greater than 88 inches	One handrail on each enclosed side and one intermediate handrail located in the middle of the stair	One stair rail system with handrail on open side, one handrail on enclosed side, and one intermediate handrail located in the middle of the stair	One stair rail system with handrail on each open side and one intermediate handrail located in the middle of the stair	
Exterior stairs less than 44 inches				One handrail on least one side

70Chapter 4

Figure 4.14   Stairs Missing Handrail on One Side.

the width of the stair tread, ship stairs and alternating tread-type stairs must always be equipped with a handrail on both sides. Figure 4.14 shows a stairway with a missing handrail.

For a better understanding of the chart from 1910.28(b)(11)(ii), it must be noted that handrails and stair rail systems are both different. A handrail is "a rail used to provide employees with a handhold for support" while traveling on stairs and is seen in figure 4.15. A stair rail system or stair rail is a "barrier erected along the exposed or open side of stairways to prevent employees from falling to a lower level" and is seen in figure 4.16. The application of a stair rail is similar to the use of a guardrail, where it prevents employees from falling off the open side of the stairs. Handrails can be part of a stair rail system, but handrails are not stair rail systems on their own. As identified in Table 4.6, if one or both sides of the stairs are enclosed, stair rail systems may not be required, and only handrails could create a safe and compliant stairway. This is because the enclosed side will prevent an employee from falling off the stairs.

Handrails and stair rail systems each have specific requirements based on the height from the stair treads. Handrails must not be less than 30 inches and not more than 38 inches as measured from the leading edge of the stair tread to the top surface of the handrail. Other criteria also apply to handrails. For example, the handrail must be a smooth surface to protect against cuts and lacerations. The design of the handrail must also be of size, shape, and dimension to allow for employees to firmly grasp the handrail when traveling the steps.

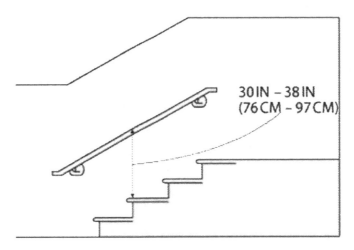

30 IN – 38 IN
(76 CM – 97 CM)

Figure 4.15 Handrail Diagram from Figure D-12 in 1910.29.

Height requirements for stair rail systems are dependent on the date of installation of the stair rails, similar to doors or gates leading to platforms from stairs. For stair rails installed before January 17, 2017, the height of the top of the stair rail system is not permitted to be less than 30 from the leading edge of the stair tread to the top surface of the top rail. Stair rails installed on

Figure 4.16 Stair Rail System Installed on or after January 17, 2017.

or after January 17, 2017, must not be less than 42 inches from the leading edge of the stair tread to the top surface of the top rail. The top rail of a stair rail system may be used as a handrail only when the height of the stair rail system is not less than 36 inches and not more than 38 inches as measured at the leading edge of the stair tread to the top surface of the top rail.

Any openings in stair rail systems must not exceed a maximum of 19 inches at its least dimension. Keeping openings less than 19 inches help to eliminate falls through stair rail systems. Similar to the top rail in a guardrail system, handrails and the top rail of a stair rail system both must have the ability to withstand a force of at least 200 pounds applied in any direction.

Ladders and Fixed Ladders

One of the most commonly used tools in occupational settings is a ladder, either portable or fixed. Workers frequently use ladders to increase working height and to access upper levels and platforms. In scenarios where stairs are either not installed or are not feasible to use, ladders provide an effective alternative means of access.

Portable ladders come in a variety of styles and sizes. Step ladders (figure 4.17) are designed only for use as a working platform. They are not intended to be used to access another platform. A commonly observed unsafe use of a step ladder is when employees fold a ladder and lean it against a surface to be closer to the face of their work. Employees also commonly fold and lean ladders in this manner and then climb the rungs to access an upper level or platform. Neither of these methods of use is safe, and they do not meet the manufacturer's provisions for safe use.

Other key safety items to recognize when using a step ladder include the use of the cap of the step ladder and top step. Neither surface is intended to be used as a working surface or a step. The maximum height a user is permitted to climb to on a portable step ladder is two rungs down from the cap of the ladder. This is marked on the ladder with appropriate warning statements.

Portable step ladders must be equipped with a metal spreader or locking device that securely holds the front and back sections of the step ladder in an open position while the ladder is in use. Damage to the spreader or locking device may result in the ladder being unstable.

In scenarios where portable step ladders are used on slippery surfaces, the ladder must be secured to prevent accidental displacement. Ladder rungs must be intact and free of substances that could be slippery and lead to falls such as oils, greases, snow, and ice. If employees are working on portable step ladders in a doorway, aisleway, or passageway, steps must be taken to protect the worker on the ladder. For example, a barricaded work zone is an

Figure 4.17   Portable Step Ladder.

acceptable control measure; or a spotter can be used to guide traffic away from the work area or the doorway where the ladder is placed.

Platform ladders (figure 4.18) are similar in their use to a portable step ladder. However, unlike step ladders, platform ladders have a large top step, or platform, that employees are permitted to access to increase their working height. This style of the ladder is considered to be a safer alternative than a traditional step ladder. The increased size of the working platform helps to prevent accidental falls from the ladder and is also ergonomically safer for employees to use. Individuals who work on ladders extensively tend to suffer ergonomic and muscular-skeletal disorders of the feet and lower back. Because the large platform can help provide the employee with more balance and a better posture, it helps to mitigate some ergonomic hazards.

Another style of the portable ladder is a straight ladder as seen in figure 4.19. Some styles of straight ladders can be fastened together to make a taller ladder. When manufacturers design these features into straight ladders, industry terminology refers to them as extension ladders. These are ladders that are primarily used for accessing upper levels. When these ladders are used to gain access to an upper level or platform, the top of the ladder must exceed

Figure 4.18   Portable Platform Ladder.

a minimum of 3 feet beyond the surface of the platform being accessed. The ladder must also be secured to prevent accidental displacement during climbing.

To properly install an extension ladder for access, employers should train employees as part of their facility's ladder safety program in the 4 to 1 rule. This helps ladder users determine the proper distance away from the building that the bottom of the ladder must be placed in comparison to the height at which it is resting against the wall or building. The ladder should be ¼ of the distance away from the building or wall that the ladder rests against. See figure 4.20. For example, let's consider a straight or extension ladder that is 20 feet long and the platform being accessed is 16 feet in height. The key number to consider in this calculation using this example is 16 feet. Despite the ladder being 20 feet in length, the ladder is resting against the platform at a height of 16 feet above the ground. Divide this number by four to get 4 feet. This ladder must be placed 4 feet away from the wall of the platform being accessed. This simple calculation ensures that the installation and use of a straight or extension ladder are always installed and used at the correct and safe ladder angle.

Figure 4.19 Portable Straight/Extension Ladder.

Fixed ladders are used as permanently installed points of access and are not removable. These are commonly used in facilities to access roofs, platforms, mezzanines, and loading docks. Fixed ladders have specific requirements

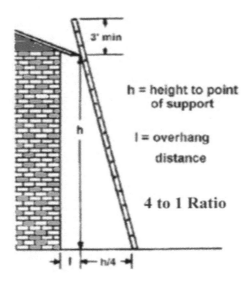

Figure 4.20 Proper Straight or Extension Ladder Angle.

based on the design and manner in which they are installed for access onto a surface or level. Similar to portable ladders that are required to extend at least 3 feet above a surface for access, the side rails of fixed ladders must extend 42 inches above the walking working surface being accessed. The side rails allow the individual using the ladder to pull themselves up from the top rung and through the ladderway when accessing another surface.

Prior to the OSHA's walking-working surfaces regulation being put into effect, there was no requirement for fall arrest or climbing systems for employees to use when climbing on fixed ladders. These fixed ladders commonly included a cage or well surrounding the outside of the ladder, but employers were never required to install any other protective systems. For fixed ladders that extend beyond 24 feet, these requirements now exist.

Fixed ladders greater than 24 feet that were installed on or after November 19, 2018, must now be equipped with a PFAS or ladder safety system (figure 4.21). These same ladders that may have been installed prior to this effective date can still use a cage or well around the outside of the ladder in lieu of a personal fall arrest or ladder safety system. However, all fixed ladders must be in compliance with new requirements by November 18, 2036. Employers should also be aware that OSHA calls for fixed ladders to be updated to reach

Figure 4.21   Ladder Safety System.

compliance with 1910 Subpart D at any point after November 18, 2018, when repairs, replacements, or alterations to the fixed ladder occurs.

PFAS used for ascending and descending the ladder would be anchored to a suitable anchorage point at the top of the ladder. Selected connecting devices and other equipment used as part of the fall arrest system must meet the requirements which were discussed earlier in this chapter.

Ladder safety systems are more commonly used with ladders taller than 24 feet, rather than PFAS. These systems are affixed to a ladder and act as a form of positioning and fall restraint that involve the use of travel restraints. Employees wear either a body belt or a full-body harness that has a front D-ring to use as a connection point for the travel restraint or positioning system. A cable is affixed to the ladder with an attached anchorage device. Users attach a short connecting device to an adjustable device on the cable prior to climbing the ladder. The attached anchorage device on the cable moves up and down on the ladder during ascent and descent. This device will stop and lock in place immediately if an employee loses hold of the ladder while climbing and falls, thus eliminating the potential for a freefall.

The cable used as part of a ladder safety system is manufactured to have a minimum tensile strength of 5,000 pounds. The cable is also drop tested with a 500 pound weight that is dropped 18 inches from the cable's anchorage device.

Regardless of the style of ladder, portable or fixed, it is imperative that the ladders not be overloaded. Fixed ladders must be manufactured and installed to support the maximum intended load. Manufacturers of portable ladders design ladders with different weight ratings and capacities, which are intended for different uses. Although OSHA does not require a minimum weight rating for a portable ladder, employers must ensure that the maximum intended load does not exceed the listed capacity of the ladder.

The maximum intended load incorporates all weight that will be applied to a ladder. For instance, the weight of the employee, tools, equipment, and materials are all part of the maximum intended load calculation.

Ladders must also be inspected prior to each use. When structural damage is identified with rungs, side rails, bracing, bases, and other structural members, the ladder must be tagged and removed from service. Fixed ladders can be repaired depending on the level of damage. However, employees must never attempt to fix or repair portable ladders. The ladder manufacturer must be contacted prior to performing any repairs to the ladder.

Similar to stairs requiring uniformed riser heights and tread depths, uniformed spacing is required on ladder rungs, regardless of the type or style. Spacing of ladder rungs must be uniform on the ladder and between 10 and 14 inches.

Access Using Fixed Ladders, Roof
Access, Ladderways, and Holes

OSHA defines a hole as "a gap or open space in a floor, roof, horizontal walking working surface, or similar surface that is at least 2 inches in its least dimension." These holes must be protected from employees accidentally falling through or from material falling through the opening onto employees working below.

When holes are identified in the floor which meet OSHA's definition, employers have a few options to protect employees. First, hole covers can be installed. These covers must be secured and capable of supporting twice the maximum intended load. Hole covers must also be marked to identify that a hole is present beneath the cover. Commonly, hole covers will be marked with the word "HOLE" written on them in orange marking paint. Covers installed in this manner will prevent employees from falling through the hole if it was large enough and will also prevent material from potentially falling onto employees below.

Employers may also choose to utilize conventional fall protection such as guardrail systems, safety nets, or personal arrest to protect employees working around open holes if there is a potential for an employee to fall through the hole. If conventional fall protection is utilized, measures must still be taken to prevent material from falling through onto employees working below.

Many facilities have skylights on their roof. Although they are covered, skylights (figure 4.22) must be treated as a hole or an unprotected edge. Unless specified otherwise, the plastic dome covers on typical skylights are not designed or engineered to support an individual's body weight from falling or stepping onto the skylight. Options to protect skylights include the use of cages or safety nets on top of the skylights. These systems can be customized to fit the style and size of the skylight. Employers may also

Figure 4.22   Unprotected Skylights.

provide PFAS or install guardrails around the skylight rather than using cages or safety nets. In some cases, the top of skylights may be constructed of glass panels. Employers should not assume that these will not shatter if an employee were to step or fall onto them until they have obtained engineered data or product specifications stating safety information on the glass panels.

Employees may be required to access a roof or another surface through a floor hole. Figures 4.23 and 4.24 show examples of a roof hatch with and without guardrails. For example, employees are often required to access roofs through a roof hatch and hatchway that is connected to a fixed ladder or stairway. While following requirements for fixed ladders and stairways listed in previous sections, employers should also recognize that additional regulations exist for access onto roofs and other platforms using fixed ladders and hinged floor-hole covers (e.g., roof hatches).

When hinged floor-hole covers such as a roof hatch as part of a hatchway are used to access a surface using a fixed ladder or stairway, employers are required to install a fixed guardrail system around all open sides of the roof hatch, with the exception of the side that is used for access. On the open side that is used for access, the employer must ensure that the hinged floor-hole cover remains closed after employee access or a self-closing gate is installed.

Figure 4.23   Roof Hatch without Guardrails or Self-Closing Gate.

Figure 4.24 Properly Protected Roof Hatch.

Either control method will prevent accidental falls through the floor hole, but the installation of a gate is more effective because it will rely less on ensuring that employees are following safe working procedures. This requirement is in place for these types of holes regardless of whether the employee accesses the level using a fixed ladder or a stairway.

Other fixed ladders used for access to platforms and upper levels through ladderways, not through hatchways or floor holes, must have protection to prevent falls. Figures 4.25 and 4.26 show ladderways that are protected vs. unprotected. Ladderways are the pass-through openings through the side rails that extend above the top ladder rung on a fixed ladder when accessing a surface. Ladderways with a height of 4 feet or more above a lower level must be protected from accidental falls through the ladderway by using either a self-closing gate that swings outward or an offset.

Training Requirements

Before any employee is exposed to a fall hazard, they are required to be trained in accordance with 1910.30, which contains training requirements as part of OSHA's walking-working surfaces regulation. Topics to include in the training include care, use, inspection, and storage of personal fall arrest equipment. As part of the training, it is imperative that the employer train

Figure 4.25   Unprotected Fixed Ladder Opening.

Figure 4.26   Fixed Ladder with Self-Closing Gate.

employees exposed to fall hazards in the proper setup and installation of a PFAS, including the installation of anchorage devices and tie-off or connecting methods. Fall protection training must also include discussions on ladders. The training should communicate the use of ladder safety systems, inspections of ladders, and the correct use of portable ladders.

Employees must also be trained in the regulatory requirements of which they are affected under 1910 Subpart D. For example, employees must understand the height requirements at which they are required to use fall protection. Employers must train their employees who are exposed to falls in the nature of fall hazards in their work areas, how to recognize fall hazards, and the steps that must be taken to mitigate fall hazards.

Employers should be customizing training to identify facility or company-specific protocols of which the employee must be aware. Providing contact personnel in the facility for the employee if questions regarding fall protection arise is important. Such personnel may include a facility engineer who assists in the process of identifying suitable surfaces for anchorage. Providing contact information for the facility's safety professional is also important as part of the training. The training must also include a review of any specific work rules or the use of a designated area for temporary and infrequent work on roofs.

Although it is imperative to train employees exposed to a fall upon hire and before they are exposed to fall hazards, training is not intended to only be provided initially or upon hire. OSHA requires that when inadequacies in knowledge are observed or changes in the workplace take place, the employer must provide the employee with refresher training.

All training must be documented. Failure to have documentation of employee training would render the employee "untrained" during an OSHA inspection. Additional or refresher training should also be documented in addition to initial training.

REVIEW QUESTIONS

1. At what height are employers required to provide fall protection to employees in general industry facilities?
2. What is the required width that must be maintained in exit routes and is a best practice for the spacing of walkways in walking-working surfaces?
3. Explain the differences between conventional and alternative methods of fall protection.
4. List the height requirements and minimum supporting weight requirements for a guardrail system.
5. What is the maximum freefall and deceleration distances that an employee can experience in a fall?
6. Equipment used in a PFAS is intended to limit fall forces that are exerted on the employee to how many pounds?

7. List three work tasks that would be considered temporary and infrequent.
8. When is it acceptable to use ship stairs, spiral stairs, or alternating tread-type stairs if regular and routine access between walking-working surfaces is required?
9. What is the minimum width and depth of a landing or platform when stepping off stairs?
10. Employees are attempting to use an extension ladder to access a platform that is 22 feet above the ground. How tall must the ladder extend and what distance away from the face of the wall must the bottom of the ladder be positioned?
11. What is the uniformed spacing that is required for ladder rungs?

REFERENCES

Occupational Safety & Health Administration [OSHA]. (2016). Regulations (Standards-29 CFR 1910.21). Retrieved from https://www.osha.gov/laws-regs/regulations/standardnumber/1910/1910.21

Occupational Safety & Health Administration [OSHA]. (2016). Regulations (Standards-29 CFR 1910.22). Retrieved from https://www.osha.gov/laws-regs/regulations/standardnumber/1910/1910.22

Occupational Safety & Health Administration [OSHA]. (2019). Regulations (Standards-29 CFR 1910.23). Retrieved from https://www.osha.gov/laws-regs/regulations/standardnumber/1910/1910.23

Occupational Safety & Health Administration [OSHA]. (2019). Regulations (Standards-29 CFR 1910.25). Retrieved from https://www.osha.gov/laws-regs/regulations/standardnumber/1910/1910.25

Occupational Safety & Health Administration [OSHA]. (2016). Regulations (Standards-29 CFR 1910.28). Retrieved from https://www.osha.gov/laws-regs/regulations/standardnumber/1910/1910.28

Occupational Safety & Health Administration [OSHA]. (2019). Regulations (Standards-29 CFR 1910.29). Retrieved from https://www.osha.gov/laws-regs/regulations/standardnumber/1910/1910.29

Occupational Safety & Health Administration [OSHA]. (2016). Regulations (Standards-29 CFR 1910.30). Retrieved from https://www.osha.gov/laws-regs/regulations/standardnumber/1910/1910.30

Occupational Safety & Health Administration [OSHA]. (2019). Regulations (Standards-29 CFR 1910.140). Retrieved from https://www.osha.gov/laws-regs/regulations/standardnumber/1910/1910.140

Occupational Safety & Health Administration [OSHA]. (1995). Regulations (Standards-29 CFR 1926.502). Retrieved from https://www.osha.gov/laws-regs/regulations/standardnumber/1926/1926.502

Chapter 5

Exit Routes and Emergency Planning

INTRODUCTION AND SCOPE

Exit routes are important for safety in the event of a fire or emergency. All workplaces must have established exit routes for employees to egress (escape, exit) in a safe manner. Exit routes are comprised of three specific parts: the exit access, the exit, and the exit discharge. Exit routes cannot be blocked. All doors leading to an exit must remain unlocked. Exit signs need to be in place when employees may need to know which direction to egress. Exit routes have to be wide enough to let people escape, and employers have to keep them free from flammable or hazardous material. Exit routes are typically recognized in an employer's Emergency Action Plan.

In addition to having exit routes for people to escape during a fire or emergency, employers must create and maintain an Emergency Action Plan. An Emergency Action Plan outlines procedures to be followed by employees in the event of an emergency. Employees need to know what to do, where to go, how to escape, and specific duties that need to be fulfilled. Related to fire emergencies, and specifically required for some fire hazards and equipment, employers must maintain a Fire Prevention Plan. This plan outlines how employers will control ignition sources and prevent fires from starting. Maintaining exit routes and emergency planning are crucial components to keeping employees safe in times of crisis.

The OSHA standards covered in this chapter include:

- 29 CFR 1910 Subpart E—Exit Routes and Emergency Planning
  - 29 CFR 1910.34—Coverage and definitions
  - 29 CFR 1910.36—Design and construction requirements for exit routes

- 29 CFR 1910.37—Maintenance, safeguards, and operational features for exit routes
- 29 CFR 1910.38—Emergency action plans
- 29 CFR 1910.39—Fire prevention plans

NARRATIVE

Margot is a working mother, and the year is 1955. Margot works in a textile factory in New York City making clothes. Her husband is disabled, so she works up to 70 hours a week at the factory to provide for her three children. The textile factory is not an ideal place to work.

The conditions of the factory where Margot works are brutal. The management only cares about production and not the employees' health and safety. Recently, management decided to crack down on distractions. Management does not want their employees making phone calls, taking smoke breaks, or frankly doing anything other than working. During the long, grueling workday, management locks the exit doors of the factory floor. So even to use the bathroom, Margot has to ask permission to have a door unlocked, and she is timed until she returns to her station. The factory is hot and miserable, and management has turned the workplace into a cruel labor dungeon.

When Margot comes home from her work dungeon, she only has a few hours to spend with her family before she has to go to bed and do the workday over again. Those few hours are the only time of joy in Margot's life. Though, even when she's home, Margot is working to take care of her family. Margot's three boys rejoice when she is home. The young children love and depend on their mother. If anything were to happen to Margot, those boys' world would fall apart.

It is a Saturday, and as usual, Margot has to report to the factory for work. She puts her head down and does her job. Nine hours into her shift, Margot finally is approaching the end of her workday. At about 4 pm, one of the other employees returns from her usual, secret smoke break in a broom closet. She quickly snuffs the cigarette on her way back on to the factory floor and tosses it in a bin of scrap cloth material. Within a few minutes, the partially extinguished cigarette sets the bin of cloth on fire. Factory workers notice the smell and find the flame, but there are no fire extinguishers to be found. The flame is spreading at an alarming rate given the poor conditions of the factory. Surrounding cardboard boxes and pieces of clothing quickly ignite and the flames spread.

As the flames spread quickly, smoke fills the production floor on the eighth story of the factory building. There are no management employees around. All management employees are on the upper floor. No phones are around to call up to management for help. Factory employees run to open windows and let the smoke out. They run to the exit doors in hopes to escape only to be reminded the doors are locked and chained shut. Employees are in sheer panic banging on doors and screaming from open windows. With every minute that passes, the flames spread and grow hotter. Smoke is thick. Margot is crouched on the floor near a window gasping for air mostly inhaling smoke. Margot is thinking about her children. She can hear their voices in her head clearly despite the screaming and the banging on doors. As the flames grow to unbearable intensities, some employees are trapped in the interior of the factory floor. They scream in agony as they are burning alive. Other employees, like Margot, are now trapped between the flames and the exterior wall of the factory. Margot faces the window and closes her eyes as other employees continue to scream and beg for help. The back of Margot's body is burning from the heat. The front of her body feels the cool air through the window. She realizes she can choose her fate. She will either burn to death, or she can jump from the eighth story window. As her clothes burn from her back, she decides anything would be better than burning to her death. She sits on the window seal and only closes her eyes for a minute to think of her children's faces. She leans forward, holds the images of her boys' faces in her mind, and falls for a few seconds before striking the pavement below. If only the exit door were unlocked and employees were given their right to a safe escape from the fire, Margot's three children would have got to see their mother again.

SUMMARY OF OSHA STANDARDS

Exit Routes

Definitions

There are several important definitions to understand when interpreting OSHA standards about exit routes. An *exit* is a portion of an exit route that is generally separated from other areas to provide a protected way of travel to the exit discharge. An example of an exit is a two-hour fire-resistance-rated enclosed stairway that leads from the fifth floor of an office building to the outside of the building. An *exit access* is the portion of an exit route that leads to an exit. An example of exit access is a corridor (hallway) on the fifth floor of an office building that leads to a two-hour fire-resistance-rated enclosed stairway (the Exit). An *exit discharge* is the part of the exit route that leads directly outside to a safe area of refuge. An example of an exit discharge is a pathway from a door at the bottom of a two-hour fire-resistance-rated enclosed stairway that discharges to a safe place. Figure 5.1 illustrates an exit route and its components. *Occupant load* means the total number of persons that may occupy a workplace or portion of a workplace at any one time. These definitions will be helpful reading onward.

Exit Route Requirements

An exit route must be a permanent part of the workplace. It cannot be removed or blocked even during temporary construction activities. The exit (e.g., protected stairway leading from fifth floor to ground floor) must be constructed of materials that have at least a one-hour fire-resistance rating. For example, most exit route stairwells are made of cinder blocks. Figure 5.2 shows a stairway exit. There must be at least two exit routes established in all workplaces. The size of the building, type or work activity, occupant load, and local fire codes may require more than two exit routes. For assistance in determining the minimum amount of exit routes in a workplace, employers are directed to consult NFPA 101-2009, Life Safety Code. This is the gold standard for exit route design requirements.

An exit route must meet minimum height and width requirements to reduce the risk of a bottle-neck situation when employees may be stampeding to escape in a fire or emergency. All points of the exit route must be at least 7 feet 6 inches high. If there are things sticking out or protruding from the ceiling, they may not restrict the height of the exit to less than 6 feet 8 inches. The width of the exit access (e.g., width of hallways to reach an exit stairwell) may not be less than 28 inches (accounting for protruding objects). All points of the entire exit route shall not be less than what is

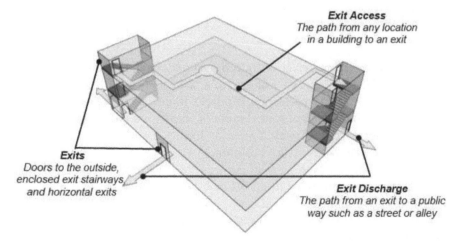

Exit Access
The path from any location in a building to an exit

Exits
Doors to the outside, enclosed exit stairways, and horizontal exits

Exit Discharge
The path from an exit to a public way such as a street or alley

Figure 5.1

Figure 5.2

required to accommodate the number of people that may use the exit route. Precise required measurement can be found using NFPA 101-2009. It is also recommended to consult your state-specific fire code when determining if components of an exit route are wide enough. Outdoor exit routes are uncommon but are allowed (e.g., fire escapes on city buildings). They must have guardrails to protect people from falling, and they must be maintained to be free from ice and snow. Outdoor exit routes may not have a dead-end that is longer than 20 feet.

In addition to requirements for parts of an exit route, doors leading to exits have OSHA requirements. Any door leading to an exit must always be unlocked and be able to be opened from the inside. These doors cannot have fancy devices that keep the door locked during normal times and unlock the door if a fire is detected, because these devices could fail. Only minor exceptions apply like in prisons. Any exit door must open as normal with hinges on the side. Figure 5.3 is a photo of a door leading to an exit.

Each exit discharge must lead directly outside (e.g., the door at the bottom of a stairway exit that leads to the outside). An exit discharge, most simply put, is the final door and pathway of an exit route that leads to safety. The areas where employees may congregate after egressing through an exit discharge are often referred to as assembly points or outdoor areas of refuge. This area must be safe if the building is on fire, and it must be able to fit the number of people that might congregate in the area.

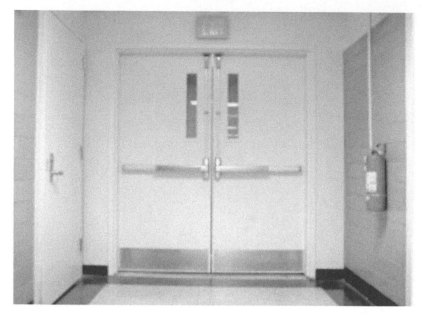

Figure 5.3

Exit Route Safeguards and Maintenance

Exit routes must be maintained to provide protection in the event of a fire or emergency. Exit routes must be kept free of flammable, ignitable, and combustible substances. So, the employer cannot store cardboard boxes and flammable substances in exit route access (e.g., hallway, corridor) or exits (e.g., stairwells). Exit routes shall be free from obstructions or materials that may fully or even partially block the path of egress. Any safeguards that are part of an exit route must be in working order at all times such as sprinkler systems, alarm systems, self-closing fire doors, and lighting. Fire doors are doors that are rated to withstand fire for a period of time. Fire doors shall not be propped open.

Each exit route door must be visibly marked by a sign reading "Exit." If the direction of travel to the exit or exit discharge is not immediately apparent, signs must be posted along the exit access indicating the direction of travel to the nearest exit and exit discharge (e.g., signs in hallways that point you in the direction to the nearest exit). So, if you are standing on a factory floor and look around, you should be able to identify which direction to travel to the closest exit. Additionally, the line of sight to an exit sign must clear and unobstructed at all times. Exit signs cannot be covered by decorations. Figure 5.4 gives an example of an exit sign posted in an exit access. If any door along an exit access could be mistaken for an exit, it must be marked "not an exit" or must be marked with a label of its actual use (e.g., closet).

Figure 5.4

Exit signs do not have to have red letters; however, there are some requirements for the sign. The sign must be illuminated during normal conditions
and when the power goes out (e.g., by an internal light source, outside light
source, self-luminous). The word "Exit" must be legible with letters that are 6
inches high and the letters themselves (each stroke of the letter, not the entire
letter) must be at least ¾ inch wide. Most of these requirements are easy to
meet, since exit signs are generally manufactured this way; however, that is
not guaranteed.

Emergency Planning

Emergency Action Plans

In addition to maintaining exit routes, all employers are required to have an
Emergency Action Plan (EAP) in place. If the employer has more than ten
employees, then the EAP needs to be written and on file. If the employer has
ten or fewer employees, the EAP can be communicated to the employees
orally. At a minimum, an EAP must include:

• Procedures for reporting a fire or emergency

- Procedures for emergency evacuation (e.g., different types of evacuation, exit route paths to travel)
- Procedures to be followed by employees who must fulfill critical duties before exiting (e.g., shutting down equipment that might explode and make the situation worse)
- Procedures to account for all employees after evacuation (e.g., roll call and attendance)
- Procedures to be followed by anyone assigned rescue or medical duties (Although, no one is required to be assigned these duties.)
- The name/job title of every employee who may be contacted to get more information about the EAP

   Employees must be trained on safe and orderly evacuation (e.g., by conducting fire drills). Each employee shall review (or be trained on) the EAP when the EAP is first developed, when employees are first assigned an EAP-related task, when the EAP changes, or when an employee's responsibility changes per the EAP.

Fire Prevention Plans

Similar to EAPs, many employers are required to have a written Fire Prevention Plan (FPP). Specific OSHA standards require an FPP based on equipment present or hazardous (flammable) substances present in the workplace. For example, if the workplace contains very flammable substances such as ethylene oxide, methylenedianiline, or 1,3-Butadiene (above certain quantities/concentrations), then the employer must have an FPP. Also, having both an EAP and an FPP can exempt employers from certain fire extinguisher requirements, if they do not want employees to use fire extinguishers. OSHA strongly recommends all workplaces have an FPP. As an S&H professional, you should advise your employer to have a written FPP regardless of hazards present. If the employer has more than ten employees, then the FPP needs to be written and on file. If they have ten or less employees, it can be communicated to the employees orally. At a minimum, an FPP must include:

- A list of all major fire hazards, proper handling and storage procedures for hazardous materials, potential ignition sources and their control, and the type of fire protection equipment necessary to control each major hazard (e.g., types of fire extinguishers)
- Procedures to control accumulations of flammable and combustible waste materials
- Procedures for regular maintenance of safeguards installed on heat-producing equipment to prevent the accidental ignition of combustible materials

- The name or job title of employees responsible for maintaining equipment to prevent or control sources of ignition or fires
- The name or job title of employees responsible for the control of fuel source hazards

An employer must inform employees when they start their job of the fire hazards to which they are exposed. An employer must also review with each employee those parts of the FPP necessary for self-protection. Because of the overlap of required procedures and related applications, EAPs and FPPs are often combined into one written program.

REVIEW QUESTIONS

1. Describe the three components of an exit route. Give an example of each.
2. What is the minimum height of an exit route?
3. What is the minimum width of an exit access per OSHA? What reference would you use to find the minimum width of all parts of an exit route based on occupancy?
4. When is it appropriate to obstruct or block an exit route? What is not allowed to be stored in an exit route? What if you have shelving or similar items in an exit route, but it does not restrict the width to less than its minimum required dimension?
5. Where must exit signs be posted if the direction to exit is not immediately apparent? What are some requirements regarding exit signs? Can an exit sign be green?
6. Describe what procedures are required to be part of an EAP?
7. Describe what procedures are required to be part of an FPP?

REFERENCES

Occupational Safety & Health Administration [OSHA]. (2011). Regulations (Standards-29 CFR 1910.34). Retrieved from https://www.osha.gov/laws-regs/regulations/standardnumber/1910/1910.34

Occupational Safety & Health Administration [OSHA]. (2014). Regulations (Standards-29 CFR 1910.36). Retrieved from https://www.osha.gov/laws-regs/regulations/standardnumber/1910/1910.36

Occupational Safety & Health Administration [OSHA]. (2002). Regulations (Standards-29 CFR 1910.37). Retrieved from https://www.osha.gov/laws-regs/regulations/standardnumber/1910/1910.37

Chapter 5

Occupational Safety & Health Administration [OSHA]. (2002). Regulations (Standards-29 CFR 1910.38). Retrieved from https://www.osha.gov/laws-regs/regulations/standardnumber/1910/1910.38
Occupational Safety & Health Administration [OSHA]. (2002). Regulations (Standards-29 CFR 1910.39). Retrieved from https://www.osha.gov/laws-regs/regulations/standardnumber/1910/1910.39

Chapter 6

Occupational Noise Exposure

INTRODUCTION AND SCOPE

Most people do not think of noise as a hazard unless it is loud enough to hurt their ears. But, noise does not have to hurt to cause hearing loss. Even noise levels that you may consider moderate can cause hearing damage and hearing loss over time. This inconspicuous, insidious, latent effect of hearing loss is exactly why it is so important to protect employees from hazardous noise. Noise-induced hearing loss (NIHL) happens over time, and you do not realize the effect as it is happening. If exposed to hazardous noise, day in and day out, for many years, NIHL can have a major impact on quality of life. Hearing loss is irreversible, and it can make communicating with friends and loved ones difficult. Hazardous noise in the workplace must be identified, measured, and then controlled to prevent NIHL.

Industrial environments can be very noisy. Machines, equipment, and work processes all contribute to noise levels. To clarify, noise is a cluttered combination of various sounds; however, usually the words noise and sound are used interchangeably. As a rule of thumb, if you have to raise your voice to talk to someone while standing an arm's length away, the environment probably contains levels of noise that can cause NIHL over time. However, the amplitude of sound is only half of the problem. The duration of exposure is just as important. If exposed to relatively loud noise for just a short time, hearing can recover. For example, if you attend a rock concert for a couple of hours, you might notice short-term hearing loss. But, if you experience high levels of noise all day every day, you can experience permanent hearing loss.

The pathology (cause and effect) of NIHL involves some intricate physics and anatomy, but it can be concisely summarized. A sound source (like a speaker playing music) vibrates at various rates and amplitudes creating

95

sound pressure waves. A sound or noise source can be anything that creates these sound pressure waves (e.g., vibrating equipment and machinery). These waves create compressions (pushing together) and rarefactions (spreading apart) of particles in the air. These sound pressure waves travel through the air as acoustical energy and reach a human ear. The outer ear funnels the sound wave to the tympanic membrane (eardrum). The eardrum physically moves the three smallest bones in the human body that are linked in series and located behind the eardrum, in the middle ear. The acoustical energy is converted to mechanical energy conducted by the bones in the middle ear. The last bone in the series of the middle ear vibrates the oval window of the cochlea located in the inner ear. This transforms the mechanical energy into fluid energy. The fluid waves travel through the cochlea and apply various pressures onto a membrane which stimulate cilia hair cells. These tiny hair cells, which make up the organ of Corti, transmit the energy as an electrical signal to the brain. Our brain then interprets the various signals as characteristics of sound. Figure 6.1 shows a simple diagram of the human ear.

When it comes to NIHL, the key part of this energy transmission process is the cilia hair cells. If exposed to loud, high-energy noises for long durations, the sound waves and associated pressure damage these hair cells. Damage and eventual death of cilia hair cells result in permanent hearing loss. When it comes to short-term or low-energy waves, hair cells recover just fine. But long-term, high-energy, repeated exposure bends the hair cells to a point of breaking (or at least non-recovery). This effect can be compared to walking on grass. If you walk on blades of grass once or every once in a while, the blade of grass recovers from its tramped-on, bent-over position. But, if many people walk the same path of grass every day, the grass cannot recover and dies.

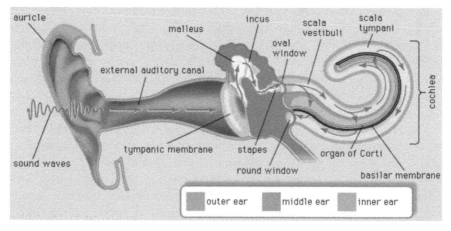

Figure 6.1

The OSHA standard on occupational noise is not overly complicated, but there are some technical concepts and mathematics that need to be understood. This chapter briefly introduces the basic physics of sound and the calculations required to evaluate noise exposure. The primary purpose of the OSHA standard on occupational noise is to set a permissible exposure limit (PEL) and action level (AL) for noise. If noise levels are measured to be equal to or exceed the PEL, employers must implement engineering or administrative controls to reduce the risk of NIHL. If these controls are not enough to reduce noise to safe levels, then hearing protection is required for employees that are exposed to such levels. The PEL for noise depends on the measured noise levels and duration of exposure to those levels. If there is potential that noise levels may exceed the AL, then the employer must have a noise level monitoring program to assess the hazard. If noise levels are determined to equal or exceed the AL, then employers must implement a Hearing Conservation Program (HCP).

An HCP includes procedures for preventing hearing loss. It requires regular noise monitoring. Those employees with exposures that are covered under the HCP must receive annual audiograms (hearing tests) to determine if they are experiencing NIHL at work. Also, employees under the HCP must be provided hearing protection and receive annual training covering the hazards of noise and the procedures within the HCP.

The OSHA standards covered in this chapter include:

* 29 CFR 1910.95—Occupational Noise Exposure

  Not included:

* Under 29 CFR 1910.95:
  ○ Audiometer instrument and calibration requirements
  ○ Audiometric test and test room requirements

NARRATIVE

James worked in the sawmill from the year 1944 to the year 2001. He retired at age seventy-five. The sawmill has huge saws and machinery that run all day every day. These saws cut slabs of timber from trees into usable pieces of lumber. They are extremely loud. The mill is out in the middle of nowhere, so there are not any neighbors bothered by the sound. But inside, the noise levels consistently exceed 90 decibels (hazardous for relatively short exposure times). The mill is a small place and a family-owned company. The owner is not educated about occupational safety and health and has never even thought about the hazard of noise. The mill does happen to have a dirty coffee can half-filled with earplugs by the entrance, but most employees who work there do not even know it is there, nor would they care to use earplugs. James was just as naive to the hazard of noise as the rest of the employees at the mill. He came to work each day, walked straight past the can of earplugs, cut timber, and did his job.

In addition to the hazardous noise that James experienced at work, eight hours a day, for more than fifty years, James had noisy hobbies his whole life. In his younger years, James was an avid motorcycle rider. He enjoyed working in his personal wood shop at home. Not only was he exposed to hazardous noise while on the job, but his hearing was also damaged in the evenings and on weekends. Over the decades of working in the noisy mill and practicing his loud hobbies, James slowly lost his hearing. He did not notice right away, or even day by day, but the delicate hair cells within his inner ear were damaged or destroyed, little by little, each day of his life. Over time, these hair cells do not repair.

James' hearing loss progressed as he aged. By the age of fifty, James' hearing loss caused tension in his marriage. Every time James' wife tried to talk to him, James had to ask her to repeat herself. Understandably, this was incredibly frustrating. By the age of sixty, James was a granddad. He looked forward to each visit with his granddaughter, but he struggled to hear her small voice and laughter.

Now, in his seventies, the effect of NIHL has compounded with natural hearing loss due to age. This has rendered James nearly deaf. Hearing aids do not do James any good. His delicate sense of hearing has been practically destroyed. James does not even bother to try and talk with his granddaughter anymore, who at this point is old enough to tell him about her day and her time at school. He cannot hear or understand a word she says. James spends most of the day in his armchair with the muffled rumble of life events passing him by. He cannot help but be bitter and wish he knew better about the effects of noise on his hearing and how losing his hearing would so drastically affect his life. While the sight of his granddaughter's smile makes him happy, it breaks his heart that he cannot hear her joyous laughter.

SUMMARY OF OSHA STANDARD

Technical Background Information and OSHA Exposure Limits

The PEL for exposure to continuous noise is 90 dB(A) as an eight-hour time-weighted average. Let's break that down:

The PEL is the maximum amount of noise energy (in form of sound pressure level) that employees can be exposed to as calculated as an eight-hour

TWA (time-weighted average). The PEL for noise, like most chemical and health hazard PELs, is typically expressed as an eight-hour average. This means it is not an instantaneous measurement that can be captured at any moment, but it is an average of exposure to noise over time. The PEL can be applied to exposure times that are not eight hours.

Continuous noise means that the maximum amplitudes of the sound pressure waves are at intervals of one second or less. Simply put, if you continuously hear the noise, it is continuous noise. But, if noises or sounds are individual bangs or bursts that are more than one second apart, then it is not considered continuous noise, and this PEL would not apply.

dB stands for decibels. The decibel is the unit for sound pressure level. Noise or sound is pressure through the air that makes air particles compress and decompress. These pressures are very small and measured in units that are expressed as force over an area like Newtons per square meters (N/m^2). The range of sound pressures that humans can experience can span from about 0.00002 N/m^2 (threshold of hearing) to 200,000 N/m^2 (threshold of discomfort) and above. This is a huge range! So, to make this range more manageable to quantify and talk about, we apply a logarithmic scale. This log function is applied to a ratio of the measured pressure over the reference pressure. The reference pressure is about the threshold of human hearing which is considered to be 0.00002 N/m^2. It should be noted that units of pressure other than N/m^2 can be used (e.g., Pascals). The equation to calculate decibels from sound pressure is seen in equation 6.1.

$$dB = 20 \times \text{LOG} \left(\frac{MeasuredSPL}{ReferneceSPL} \right) \qquad (6.1)$$

Now, we still need to talk about the "(A)" in the PEL of 90 dB(A). When noise levels are analyzed to compare against the OSHA PEL and the hearing conservation standard, measurements must be taken on the A-weighted scale. As you can imagine, instruments and devices that measure noise do not perceive sound in the same way the human ear does. An ear is not an electronic device. The human ear is more sensitive to midrange frequencies. Frequency is like pitch in music. It is the amount of times per second that a sound reaches its maximum amplitude. The unit is expressed as Hertz (Hz). At the low end of human perception, we can hear 20 Hz. At the high end, we can hear about 20,000 Hz. The midrange frequencies are between 1,000 Hz and 4,000 Hz. The human ear is extra-sensitive to these midrange frequencies and less sensitive to the very-high and very-low frequencies. To compensate for this, instruments need to measure noise in the way the human ear perceives it. That means instruments have to amplify the midrange frequencies and turn down the amplitude on high and low frequencies. This increase in weighting (impact on the overall measurement) of the midrange frequencies and decrease in weighting of the high and low frequencies is called the A-weighting. There are other types of weightings, but A is used for comparing measurements to the OSHA PEL and AL.

The PEL is the exposure of noise that is not allowed in the workplace without the application of engineering, administrative controls, or PPE to reduce actual exposure to below the PEL. Engineering controls are things that control the noise by design and do not rely on people to be effective. Examples of engineering controls are noise barriers and noise dampening materials. Examples of administrative controls are job rotation procedures to limit the time of exposure to hazardous noise and procedures that help reduce the amount of noise produced. As seen in the next section, there are PELs for exposure times other than 8 hours. If employees are subjected to noise that exceeds the 8-hour PEL or PELs found in figure 6.2, engineering or administrative controls must be used to reduce the noise. If such controls fail to reduce the sound below PELs, PPE (e.g., earplugs or ear muffs) must be provided and worn that reduce exposure to below the PELs. One can extrapolate PELs for times other than eight hours knowing the noise-time exchange rate.

Noise Exchange Rate and Exposure Limits

The next concept to clarify is the noise-time exchange rate. Noise exposure is about exposure to energy. This means the amplitude of the noise plays a role, and the time of exposure is important. We already know the PEL for eight hours is 90 dB(A). The PEL for four hours of exposure is 95 dB(A). The PEL for two hours of exposure is 100 dB(A) and so on. Notice the exposure limit in average dB changes by −5 when we cut our time of exposure to half. This is because OSHA uses an exchange rate of 5 dB. This means, OSHA approximates that a reduction of 5 dB cuts the sound pressure level to half. Conversely, an increase of 5 dB approximately doubles the sound pressure level. This is due to the unit dB being a logarithmic scale. It would be wrong to say doubling decibel level doubles sound pressure level, because based on the exchange rate, it would be increased by much more than that. When we change level of exposure (dB), we have to adjust the allowable time of exposure accordingly (i.e., if sound pressure is doubled, the allowable exposure time is reduced in half). OSHA provides a chart of PELs based on this 5 dB exchange rate and extrapolates for us as shown in figure 6.2. OSHA does not allow exposure to continuous noise measured above 115 dB(A) for any amount of time (i.e., even for a few seconds).

Allowable exposure times for noise levels not seen in figure 6.2 are presented in 29 CFR 1910.94 Appendix A. They can be calculated when not

Duration per day, hours	Sound level dBA slow response
8........................	90
6........................	92
4........................	95
3........................	97
2........................	100
1 1/2	102
1........................	105
1/2	110
1/4 or less............	115

Figure 6.2

explicitly in tables in charts. Equation 6.2 shows how to calculate an allowable exposure time (T) based on a noise level (L).

$$T = \frac{8}{2^{\left(\frac{(L-90)}{5}\right)}} \tag{6.2}$$

Two other limits are set by OSHA. Continuous noise may not exceed 115 dB(A), even if it lasts for a few seconds. This is the ceiling limit. OSHA also sets a peak noise level for impact noises. Impact noises would be individual sounds that are sporadic, occur more than one second apart, or even happen just once. These loud bangs or impact noises may not exceed 140 dB(Z). Notice how the peak noise level for impact noises is in the Z-weighted scale. This scale weighs all frequencies evenly. When measuring impact noises, measurement devices must be set to the Z-scale to compare to the peak limit.

Hearing Conservation Program

An employer must implement a HCP when employee noise exposure is equal to or exceeds an eight-hour TWA sound pressure level of 85 dB(A). As we know from the OSHA exchange rate, 85 dB(A) is about half or 50 percent of the energy exposure as compared to the 90 dB(A) PEL (which is 100 percent of allowable dose). This 85 dB(A) HCP criteria is known as the AL, because at this level of exposure, preventative action must be taken to conserve employees' hearing. It is important to note that this AL measurement and the requirement to implement a HCP is without regard to any attenuation (protection and reduction) by hearing protection devices (e.g., plugs and muffs). When measurements indicate exposure at or above the AL (50% allowable dose), hearing protection devices must be available in the workplace and HCP procedures must be implemented.

Noise Monitoring or Sampling

When information or noise monitoring indicates employees' exposure may (potentially or actually) exceed the eight-hour TWA AL (85 dB(A)), the employer must do noise sampling or monitoring. The primary design of the noise sampling program is to identify those employees who need to be included (follow procedures of) the HCP and to enable the selection of proper hearing protection devices. Two devices are commonly used for noise measurement and sampling. A handheld sound level meter generally provides an instantaneous measurement of noise. Some can also integrate measurements of time to produce an average exposure. These are generally used for "spot-checks" in locations around a facility or perhaps for long-term sampling in a specific area (area sampling). A noise dosimeter measures noise over time to measure the TWA and noise dose (percent of allowable exposure), and

Figure 6.3

dosimeters are typically worn by employees during a shift to give an individual's noise exposure over time (personal sampling). Figures 6.3 and 6.4 show a sound level meter and a noise dosimeter, respectively.

When configuring noise dosimeters to compare measurements to OSHA HCP requirements and PELs, the instruments should be set based on the following:

- Range: 80–130 dB(A)
- Criterion level: 90 dB(A)
- Exchange rate: 5 dB(A)
- Weighting scale: A

The range is all the noise levels that will be included in the overall calculated TWA and dose measurement conducted by the device. 80 dB is the lower limit of the range, because OSHA considers sounds less than 80 dB to be non-hazardous. 130 dB is the upper limit, because if you have noises greater than 130 dB, you have bigger problems to solve first, and levels this

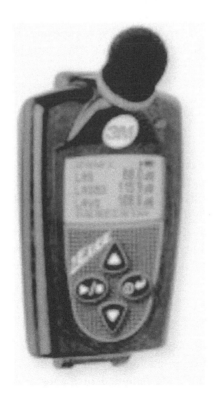

Figure 6.4

high would skew the results. OSHA does not allow continuous noise levels to exceed 115 dB(A). The criterion level is the level of noise that the instrument will recognize as 100 percent dose or maximum exposure. The PEL is 90 dB(A) and represents 100 percent of allowable dose, thus this is the value used as the criterion level. 85 dB(A) would represent 50 percent dose based on the exchange rate we discussed earlier. The A-weighting allows the instrument to perceive sound like the human ear does.

When the daily noise exposure for an employee is composed of two or more periods of noise exposure of different levels, their combined effect should be considered, rather than the individual effect of each. We use equation 6.3 to calculate allowable dose for different exposures over a time period.

$$\%\text{Dose} = \frac{C_1}{T_1} + \frac{C_2}{T_2} + \frac{C_n}{T_n} \times 100 \qquad (6.3)$$

where:

C_n indicates the total *time of exposure* at a specific noise level. T_n indicates the *time* of exposure *permitted* at that level (e.g., found in figure 6.2).

Once percent-dose (%dose) is calculated using equation 6.3, we can use that dose to calculate the eight-hour TWA exposure. Equation 6.4 is used to calculate a TWA from %dose (D). This TWA can be compared to OSHA PELs.

$$\text{TWA} = 16.61\left(\frac{D}{100}\right) + 90 \qquad (6.4)$$

Noise monitoring instruments must be calibrated according to the manufacturer's instructions to ensure accuracy. Monitoring shall be repeated whenever there is a change in the work environment, change in noise, or change in noise controls. Employees must be notified if noise measurement results indicate they have an exposure at or above the AL. Employees shall also have an opportunity to observe the noise monitoring process. Noise monitoring results shall be retained for two years.

Audiometric Testing Program for Employees in the HCP

As part of the HCP, in addition to conducting noise sampling, the employer shall establish an audiometric testing program (hearing tests) for any employees who must be included in the HCP (exposure to the AL). Employees in the HCP must receive a baseline audiogram within six months of an employee's first exposure above the AL. He/she should get this baseline test as soon as possible. Future audiograms (hearing test results) are compared against this baseline to see if the employee is losing hearing from noise exposure at work. Figure 6.5 gives an example of audiogram test results. If the employer uses a mobile test-van service for audiograms, then they have to do the baseline audiograms within a year instead of six months.

When an employee takes their baseline audiogram, they must not have exposure to workplace noise for at least fourteen hours before the hearing test. Each employee included in the HCP must receive an audiogram annually after their baseline test. Each employees' annual audiogram is compared to his or her baseline audiogram to see if a Standard Threshold Shift (STS) has occurred. An STS means that in order to hear a sound at some particular frequency that sound needs to be at a higher amplitude than it used to be for them to hear it. This means hearing loss has occurred. Another way to say it is: An STS means the employee is less sensitive to particular sounds than they used to be. Hearing loss can occur only for specific frequencies, and it is most common in midrange frequencies (since humans are most sensitive to that range). This is why hearing test consists of presenting "beeps" of single-pitch sounds at different frequencies. An STS is defined as a change in hearing

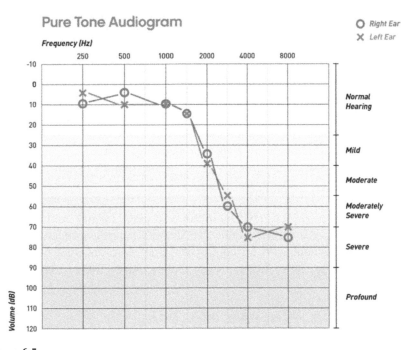

Figure 6.5

threshold relative to the baseline audiogram of an average of 10 dB or more at 2,000, 3,000, and 4,000 Hz in either ear.

It should be noted that audiogram results are corrected to account for the natural hearing loss that occurs with aging (presbycusis). Audiogram instruments and rooms must meet some specific requirements, but they are omitted from this chapter.

If an STS is observed, the employer may obtain a retest within thirty days and consider the retest valid if they choose. The audiologist, otolaryngologist (ears, nose, and throat doctor), or physician overseeing the audiograms shall determine if further evaluation of the employee's hearing is needed. If an STS is observed, the employee must be informed, in writing, within twenty-one days of the determination. Unless a physician determines the hearing loss is due to something other than work-related noise exposure, then the employer shall ensure the following steps are taken:

• Employees not using hearing protection devices are provided hearing protection, required to use them, and trained in their care and use
• Employees already using hearing protection shall be retrained on their use and provided protection with greater attenuation if needed
• Employees shall be referred to an ontological (ear) examination as needed

It is worth noting that hearing loss can be caused by noise exposure outside of work and by things other than noise. For example, certain air contaminants can cause hearing loss. Hearing loss other than NIHL is not covered in this chapter.

Audiometric test records shall include the name and job classification of the employee, date of the audiogram, the examiner name, the date of the last time the audiometer was calibrated, and the most recent noise monitoring results for the employee. Audiometric test results and these records must be retained for the duration of the employee's employment.

Hearing Protection Devices

Hearing protection devices (HPDs) in the workplace mostly are in the form of ear plugs and ear muffs. Figure 6.6 depicts hearing protectors. HPDs must be made available to all employees who have an eight-hour TWA of 85 dB(A) or greater. The employer shall require HPDs to be worn by employees who:

• May have a TWA exposure of 90 dB(A) or more
• Have not yet had a baseline audiogram but need one under the HCP
• Have had an STS

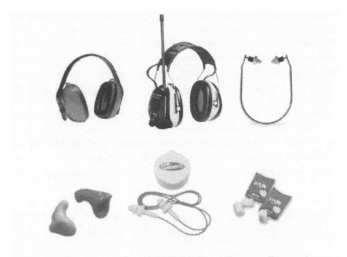

Clockwise from top left: Standard earmuffs, radio earmuffs, canal caps,

Figure 6.6

Employers must provide options of HPDs for those that require them and shall provide training on HPDs to those employees exposed to level at or above the AL.

HPDs are mandated to have a Noise Reduction Rating (NRR). This NRR is the lab-tested attenuation (level of protection) of the HPD or by how many dB it reduces the exposure level. Because these NRRs are established in lab conditions, OSHA requires employers to discount the NRR for field use. In the real world, HPDs do not fit perfectly. For A-weighted measurements, we must reduce the NRR by seven. So, if ear muffs have an NRR of thirty, their field-adjusted NRR is actually twenty-three. For an employee wearing these muff, we then can subtract the field-adjusted NRR from the measured TWA exposure. So, if the measured TWA exposure without protection was 100 dB(A), we take 100 minus 23. After applying field attenuation of the HPD, the TWA is 77 dB(A) with the HPD being worn. Employers must use this method to calculate the field NRR of HPDs. When the noise measurement is taken in a scale other than the A-scale, there are different equations used; however, these are not included in this chapter. Employers must use the field-adjusted NRR to calculate the attenuation of HPD and determine if attenuation provided by the HPD is adequate to reduce exposure to below the PEL.

Training

The employer must provide training to employees included in the HCP about the OSHA standard and procedures within the HCP. Employees must receive this training annually and must be updated with any changes that occur related to the HCP. Training shall include information on the following:

- Effects of noise on hearing
- Purpose of HPDs, advantages, disadvantages, attenuation of different types, instructions on selection, fitting, use, and care
- The purpose of audiometric tests and explanation of test procedures

Employees must be able to access training program material and the OSHA noise exposure standard must be posted in the workplace.

REVIEW QUESTIONS

1. Describe the path sound takes from a source to be transmitted to the brain. What types of energy are involved? Use anatomical terms. Why or how does hearing loss occur from hazardous noise?
2. What is the OSHA eight-hour PEL for noise? What is required of the employer to do if employees are exposed to noise levels above the PEL? What is the time allowed to be exposed to 91.53 dB(A)?
3. What is a decibel and why do we use it to measure sound? How many dB is a sound that is measured to be 0.4101 N/m²?
4. What is an exchange rate and what exchange rate does OSHA use?
5. What is the %dose of allowable noise that an employee receives when exposed to 90 dB for two hours, 83 dB for two hours, and 91 dB for four hours? What is the TWA of the employee?
6. What is the OSHA action level for noise? What does this mean? What does the employer have to do if employees are exposed to noise levels above the action level? What are the primary procedures within a HCP?
7. Describe the proper settings of a noise dosimeter if sampling noise to compare to the OSHA standard? Why is each such setting used?
8. What is a standard threshold shift (STS) regarding an audiogram? What defines it when looking at an audiogram? If an STS occurs, what does an employer have to do for the employee who experienced it?
9. What are two other ways (besides hazardous noise) that hearing loss can occur?

REFERENCES

Occupational Safety & Health Administration [OSHA]. (2008). Regulations (Standards-29 CFR 1910.95). Retrieved from https://www.osha.gov/laws-regs/regulations/standardnumber/1910/1910.95

Occupational Safety & Health Administration [OSHA]. (2008). Regulations (Standards-29 CFR 1910.95 App A). Retrieved from https://www.osha.gov/laws-regs/regulations/standardnumber/1910/1910.95

Occupational Safety & Health Administration [OSHA]. (2008). Regulations (Standards-29 CFR 1910.95 App B). Retrieved from https://www.osha.gov/laws-regs/regulations/standardnumber/1910/1910.95AppB

Chapter 7

Flammable Liquids

INTRODUCTION AND SCOPE

Flammable liquids are common substances used in manufacturing and industrial settings. But, as the name implies, they present physical hazards related to incidental ignition that can cause fires and explosions. When large quantities of flammable liquids are not used or stored properly, incidental ignition can be catastrophic. Flammable liquids can pose health hazards as well, but this chapter focuses on general OSHA requirements to prevent ignition.

A flammable liquid is classified as such based on its flashpoint. A flashpoint is a minimum temperature at which a liquid gives off vapor in sufficient concentration to form an ignitable mixture in air. It is a subtle but important distinction that it is the vapor of a flammable liquid that ignites, not the liquid itself.

This chapter summarizes and presents the key requirements of flammable liquid storage in large, permanent tanks and incidental (not part of primary processing/operations) storage and use. The OSHA standard for the flammable liquid is rather lengthy and provides rules related to the design and construction of storage vessels, as well as requirements for some specific industries and applications. These design standards and specific applications are omitted from this chapter. Only the most common, broadly applicable standards are summarized in this chapter. For more detailed information on the topics omitted from the scope of the chapter, refer to the OSHA standard on flammable liquids, especially if you work in a setting that manufactures or processes flammable liquids, if you work for an employer who designs flammable liquid vessels, or you work for an employer who maintains bulk storage of flammable liquids.

The OSHA standards covered in this chapter include:

- 29 CFR 1910.106—Flammable Liquids

Not included:

- Under 29 CFR 1910.106—Flammable Liquids
  ○ Flammable liquid tank design, construction, and materials requirements
  ○ Flammable liquid vessel piping, valves, and fittings requirements
  ○ Flammable liquid storage cabinet design, construction, and materials requirements
  ○ Flammable liquid storage requirements for dedicated storage buildings and outside storage areas
  ○ Flammable liquid requirements for mixing, drying, evaporating, filtering, distillation, or other similar operations that do not involve chemical change
  ○ Flammable liquid requirement in bulk plants and processing plants
  ○ Flammable liquid requirements in OSHA defined as "special enclosures"
  ○ Flammable liquid requirements for bulk transportation

NARRATIVE

Dave grew up, lives, and works in a small town. He works at the local plant. Dave has worked at the plant since high school. Such a thing is not unusual, because just about everyone in the town works at the plant. The small town was built around the plant, and it is the main reason people planted roots and started families where they did. The plant employs the vast majority of workers in the town. If the plant were ever to shut down, the town would collapse in more ways than one. It is the heart and soul of the rural area.

Dave works in the maintenance department of the plant, and he has been working there for thirty years. Dave has seen the plant go from a new, state-of-the-art facility for its time to a place that is behind on technology and lacking in its employee safety programs. The plant is so busy that it seems like management does not have time to keep up with OSHA requirements or even basic housekeeping for safety. The plant floor tends to be a mess with things scattered about storage areas and processing departments. The finishing department, which uses various containers of flammable liquids, is the worst when it comes to housekeeping and conditions. Since the finishing department is the last step of the manufacturing process, employees are especially rushed. There are uncovered containers of solvents and chemicals spread about the finishing department.

On a typical Tuesday, Dave is working on fabricating a replacement component for a machine located in the finishing department. Dave fabricates the part, in the machine shop, and makes his way to the finishing department to weld it in place. Without much thought, Dave puts on his welding helmet, sets up his equipment, and starts to weld. No one has ever talked to Dave about the requirements for welding and ensuring fire hazards are controlled. No one at the plant discusses fire safety, hazards related to ignition sources, or hazards of flammable liquids. Dave does not know that he is welding right next to an open container that contains a highly volatile, flammable liquid.

It does not take long for spark to jump from Dave's welding operation into the path where highly flammable vapors are evaporating from the liquid nearby. In an instant, there is a huge flash. The spark ignites the vapor, and a cloud of flames and energy are dispersed. This causes a chain reaction. There are other small and large containers of flammable liquids in the surrounding area. The vapors of these liquids are ignited, or the containers reach high enough temperatures that they rupture from the pressure, and even more flammable vapor is released turning this ignition into a catastrophe.

Dave is badly burnt and injured from the explosion. Another employee drags him out of the plant to safety. Dave survives but receives third-degree burns coving the majority of his body, and he spends months in burn recovery receiving agonizing treatments. Several employees died in the fire and the building was destroyed. Being already behind on manufacturing deadlines and barely staying economically viable, the plant cannot rebuild or recover. The plant remains shut down and is mostly a pile of burnt rubble. Not only were several employees injured or killed, but hundreds lost their job. Families struggle to feed their children and are forced to move away and look for work. The town was never the same. Now, those who have to drive through the now ghost-town may wonder what happened to the building. Those who do still live there are forced to look at the charred structure as a reminder of what used to be, what was lost, and the souls of victims who may haunt the burnt corridors that still remain.

SUMMARY OF OSHA STANDARD

Flammable Liquid Categories

A flammable liquid (FL) is any liquid that has a flashpoint at or below 199.4°F (93°C). Flammable liquids are separated into categories. Category 1 FLs have the highest flammability risk and Category 4 FLs have the lowest flammability risk; however, all categories can be ignited and should all be considered a risk related to ignition of their vapors. Category 1 FLs have flashpoints below 73.4°F (23°C) and have a boiling point at or below 95°F (35C). Category 1 FLs are less common but very dangerous. Examples include pentane and diethyl ether. Also, any flammable aerosols are considered to be Category 1 flammable liquids. Category 2 FLs have flashpoints below 73.4°F (23°C) and have a boiling point above 95°F (35°C). Examples of Category 2 FLs are acetone, gasoline, ethanol, and benzene. Category 3 FLs have flashpoints at or above 73.4°F (23°C) and at or below 140°F (60°C). Examples of Category 3 FLs include kerosene, diesel fuel, cleaning solvents, and motor oil. Category 4 FLs have flashpoints above 140°F (60°C) and at or below 199.4°F (93°C). Examples of Category 4 FLs include oil-based paints, formaldehyde, and mineral oil. Figure 7.1 illustrates the categories of FLs in a way that is easier to understand. The National Fire Protection Agency (NFPA) has a different (but similar) classification system for FLs, and safety professionals should be familiar with this, because local fire codes will have FL quantity limits based on NFPA classifications.

Outside Above-Ground Permanent Flammable Liquid Storage Vessels

Above-ground, permanent tanks containing FLs generally need to be made of steel or appropriate material and shall meet various standards for safe design. The standards for design, construction, and materials are rather lengthy and technical. They are omitted from the scope of this chapter. All above-ground FL storage tanks shall have an emergency relief vent. This device shall be able to relieve excessive internal pressure caused by heat or fires. The pressure relief device must rupture and vent in a way that does not create a hazard. Figure 7.2 shows an image of above-ground FL storage tanks within secondary containment.

The area surrounding a tank or group of tanks shall be provided with drainage or shall be diked. A dike is a wall around secondary containment to catch a potential spill. Diked areas have specific requirements such as

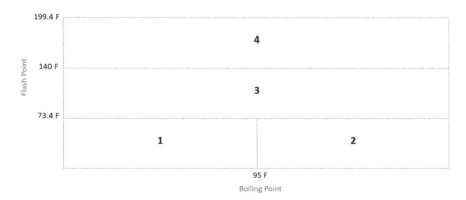

Figure 7.1

Figure 7.2

minimum volume based on the largest tank volume and other factors. The OSHA standard outlines further requirements for above-ground storage tanks and secondary containment, but these are not covered in detail in this chapter.

Permanent Underground Flammable Liquid Storage Vessels

Sometimes, FLs are stored in the bulk tanks underground. Figure 7.3 shows an image of an underground storage tank (UST) being installed. An UST containing FLs shall be at an adequate distance from property that is not owned by the employer who owns the UST. USTs storing FLs with a flashpoint below 100°F shall be no less than 1 foot from any basement or pits and shall be no less than 3 feet to another property line. The distance from any part of a UST storing flammable liquids with a flashpoint at or above 100°F to the nearest wall of any basement, pit, or property line shall be not less than 1 foot. Like above-ground tanks, USTs have many designs, construction, and material requirements omitted from the scope of this chapter.

USTs shall be set on firm foundations with at least 6 inches of materials like solid sand, earth, or gravel. Tanks shall be covered with a minimum of 2 feet of earth or shall be covered with not less than 1 foot of earth, on top of which shall be placed a slab of reinforced concrete not less than 4 inches thick. When underground tanks are, or are likely to be, subject to traffic, they shall be protected against damage from vehicles passing over them by at

Figure 7.3

least 3 feet of earth cover, or 18 inches of well-tamped earth, plus 6–8 inches of concrete (depending on the type of concrete). USTs shall be built out of corrosion-resistant material or have protection from corrosion by the use of protective coatings, protective wrappings, or cathodic protection (UST connected to a material that is sacrificed and absorbs corrosion rather than the UST corroding).

Vent pipes from USTs storing FLs with a flashpoint below 100°F shall be so located that the discharge point is outside of buildings and not less than 12 feet above ground level. Vent pipes shall discharge only upward in order to disperse vapors. Vent pipes from tanks storing FLs with a flashpoint at or above 100°F shall terminate outside of a building and shall be above normal snow level. There are many OSHA requirements about USTs that are not able to be covered in detail within this chapter.

Storage of Flammable Liquids in Containers and Portable Tanks

FLs and flammable aerosols can be stored in moveable containers, not just permanent tanks. These portable vessels can be divided into three main categories: small containers and safety cans, medium portable containers, and large portable tanks. These portable vessels have safety requirements outlined in OSHA standards. Figure 7.4 depicts an OSHA-approved "safety can." FL "medium portable containers" are typically metal drums (e.g., 55 gallon drums). Figure 7.5 depicts a "large portable tank." Portable tanks are typically large (e.g., 600 gallons) and need to be moved with a forklift. All of these vessels can be moved, either by hand or with a lifting device. They are not permanently installed like above-ground tanks and USTs. All portable containers and tanks should be approved for use by the Department

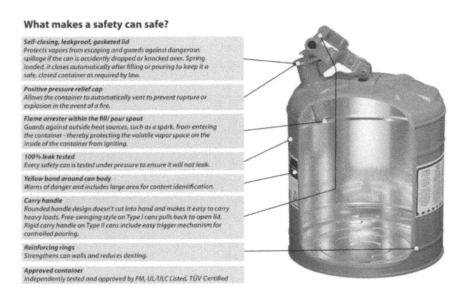

What makes a safety can safe?

Self-closing, leakproof, gasketed lid
Protects vapors from escaping and guards against dangerous spillage if the can is accidently dropped or knocked over. Spring loaded, it closes automatically after filling or pouring to keep it a safe, closed container as required by law.

Positive pressure relief cap
Allows the container to automatically vent to prevent rupture or explosion in the event of a fire.

Flame arrester within the fill/ pour spout
Guards against outside heat sources, such as a spark, from entering the container - thereby protecting the volatile vapor space on the inside of the container from igniting.

100% leak tested
Every safety can is tested under pressure to ensure it will not leak.

Yellow band around can body
Warns of danger and includes large area for content identification.

Carry handle
Rounded handle design doesn't cut into hand and makes it easy to carry heavy loads. Free-swinging style on Type I cans pulls back to open lid. Rigid carry handle on Type II cans include easy trigger mechanism for controlled pouring.

Reinforcing rings
Strengthens can walls and reduces denting.

Approved container
Independently tested and approved by FM, UL/ULC Listed, TÜV Certified

Figure 7.4

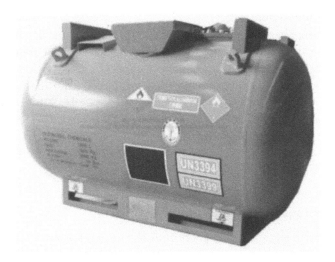

Figure 7.5

of Transportation (DOT). In other words, you should see a sticker or mark on these vessels that indicate their construction is DOT approved. This is because it is typical for these vessels to be put on a truck and transported on roadways, so the DOT sets requirements for containers to meet.

Flammable liquid containers shall be sized in accordance of Table H-12 in 29 CFR 1910.106. Container size limits depends on category and container material. For example, glass or approved plastic containers have a size limit of 1 pint for category 1 FLs and 1 quart for category 2 FLs. DOT approved metals drums have a size limit of 60 gallons. Large approved portable tanks have a size limit of 660 gallons. In office-type occupancies, storage shall be prohibited except that which is required for maintenance and operation of building and operation of equipment. Such storage shall be kept in closed metal containers stored in a storage cabinet, or in safety cans, or in an inside storage room not accessible by the public.

Safety cans have some explicit requirements that should be known. A safety can has no more than a 5-gallon capacity. It shall have a spring-closing lid and spout cover. It also must have a built-in pressure relief device, flame arrestor, and carrying handle. A flame arrestor is something that helps prevent ignition sources from reaching the vapors and helps deter vapors from escaping. Figure 7.4 describes safety can features.

Flammable Liquid Storage Cabinets

Small containers of FLs may be used in workplaces. Because FLs are a fire hazard, it is wise to store them in fire-rated cabinets to prevent flames or ignition sources from reaching them. An example of a fire-rated FL storage cabinet is seen in figure 7.6.

These storage cabinets must be constructed and have a fire resistance to limit the internal temperature to not more than 325°F. All cabinets shall be labeled with "FLAMMABLE—KEEP FIRE AWAY." Specific design and material requirements of these cabinets are omitted from this chapter. Not more than 60 gallons of Category 1, 2, or 3 FLs may be stored in a storage cabinet. Not more than 120 gallons of Category 4 FLs may be stored in a storage cabinet. For limits related to when FLs must be stored in a flammable cabinet and cannot be stored freely around a facility, refer to the section titled "Incidental Use and Storage."

Flammable Liquid Storage Room

When there are many small containers of FLs in the workplace, it is wise to store them in a specifically designed FL storage room. Figure 7.7 shows an example of an FL storage room. These storage rooms must meet design and construction

Figure 7.6

Figure 7.7

requirements that are omitted from this chapter. The maximum quantity of FL that can be stored in FL storage rooms and the maximum size of these rooms depend on if automatic fire protection (e.g., sprinkler system) is installed and the fire resistance of the construction of the room. Appendix A of this chapter provides the OSHA table that describes these requirements and limitations.

Any electrical wiring within FL storage rooms that contain FLs with flash-points below 100°F shall be approved under OSHA 29 1910 Subpart S (electrical standards) for Class 1 Division 2 hazardous locations. This means the electrical equipment must be approved for safety when flammable liquid vapors are present but are normally inside containers. All FL storage rooms must be designed with ventilation that allows for at least six air changes per hour (ACH) inside the room. If the ventilation system needs to be turned on mechanically, the switch that turns on the lights must also turn on the ventilation system. Inside storage rooms, every aisle for walking shall be maintained to be at least 3 feet wide. Containers over 30 gallons shall not be stacked on top of each other in these rooms. For limits related to when FLs must be stored in a designed FL storage room, refer to the section titled "Incidental Use and Storage."

There are also OSHA requirements for entire buildings dedicated to FL storage and bulk storage of FL containers outside of buildings. These standards are omitted from this chapter.

Flammable Liquids in General Industrial Plants

Incidental Use and Storage

OSHA provides requirements when industrial plants use FLs that are incidental to primary operations. Incidental means they are used, but not produced or used within primary processes or operations integral to production (e.g., acetone to strip paint, gasoline to fuel vehicles). All containers of flammable liquids shall be stored in closed containers or sealed tanks. FLs for incidental use must be stored in a flammable cabinet or FL storage room if exceeding these capacities:

- 25 gallons of Category 1 flammable liquids in containers
- 120 gallons of Category 2, 3, or 4 flammable liquids in containers (multiple, spread out)
- 660 gallons of Category 2, 3, or 4 flammable liquids in a single portable tank.

So, if you have more than 25 gallons of Category 1 FLs in an industrial facility that are incidental to primary operations, then the containers need to be stored in a cabinet or storage room. If you have more than 120 gallons of Category 2, 3, or 4 FLs (that are in multiple containers) in an industrial facility that are incidental to primary operations, then the containers need to be stored in a cabinet or storage room. If you have more than 660 gallons of Category 2, 3, or 4 FLs (in one single tank) in an industrial facility that are incidental to primary operations, then the tank needs to be stored in a storage room.

FLs with a flashpoint less than 100°F may only be used where there are no open flames or ignition sources that can ignite vapors. FLs shall be drawn from or transferred into vessels, containers, or portable tanks within a building only through a closed piping system. Requirements related to FL mixing, drying, evaporating, filtering, distillation, or other similar operations that do not involve chemical change are omitted from this chapter.

Sources of Ignition and Control for FL
Incidental Use and Housekeeping

Precautions must be taken to prevent the ignition of any FL vapors. Sources of ignition include but are not limited to: open flames; lightning; smoking; cutting and welding; hot surfaces; frictional heat; static, electrical, and mechanical sparks; spontaneous ignition, including heat-producing chemical reactions; and radiant heat. It's important to control static electricity when transferring FLs. Flammable liquids with a flashpoint below 100°F shall not be dispensed into containers unless the nozzle and container are electrically interconnected. This is referred to as bonding. Also, the source container shall be grounded. There must be an electrical conductor (e.g., wire) that connects the source container to something that will take incidental electrical currents to the ground. Figure 7.8 illustrates the grounding and bonding configuration.

When FLs are used and emissions of vapors are part of normal operations, electrical wiring and equipment present shall be classified as suitable for Class I, Division 1 hazardous locations according to the requirements of Subpart S (electrical standards). When FLs are stored or contained (vapors not expected to be consistently present during normal operations, but could be released only during abnormal conditions), electrical wiring and equipment present shall be classified as suitable for Class I, Division 2 hazardous locations according to the requirements of Subpart S (electrical standards). The

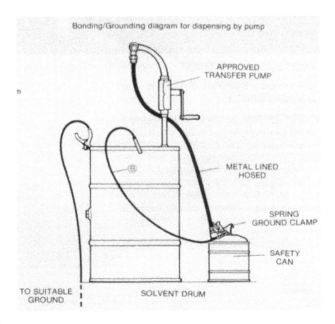

Figure 7.8

employer shall implement procedures that will control leakage and prevent the accidental release of flammable liquid vapors. Any waste and residue of FLs shall be kept to a minimum.

REVIEW QUESTIONS

1. Define a flammable liquid and discriminate between Categories 1, 2, 3, and 4 flammable liquids.
2. Name two requirements related to above-ground, permanent flammable liquid storage tanks.
3. Describe how far away from underground structures (basements) and property lines that underground storage tanks must be.
4. Describe the types of corrosion protection that are required/allowed for underground storage tanks.
5. What are the primary design requirements of an OSHA-approved safety can?
6. At what quantity do Category 1 flammable liquids that are incidental to operations need to be stored in a flammable cabinet or storage room?
7. At what quantity do Category 3 flammable liquids that are incidental to operation (stored in multiple containers) need to be stored in a flammable cabinet or storage room?
8. When is grounding and bonding required? How is this achieved? What hazard does it protect from?
9. If flammable liquid vapors are normally present in an area of a facility, what type of electrical wiring/equipment must be present? (hazardous location)
10. In a flammable liquid storage room, what type of electrical wiring/equipment must be present? (hazardous location)

REFERENCE

Occupational Safety & Health Administration [OSHA]. (2016). Regulations (Standards-29 CFR 1910.106). Retrieved from https://www.osha.gov/laws-regs/regulations/standardnumber/1910/1910.106

APPENDIX A—CAPACITY OF FLAMMABLE LIQUID STORAGE ROOMS (OSHA TABLE H-13—STORAGE INSIDE ROOMS)

Table H-13—Storage in Inside Rooms

Fire Protection[a] Provided	Fire Resistance (hours)	Maximum Size (sq. ft.)	Total Allowable Quantities (gals./sq. ft./floor area)
Yes	2	500	10
No	2	500	5
Yes	1	150	4
No	1	150	2

[a] Fire protection system shall be sprinkler, water spray, carbon dioxide, or other system.

Chapter 8

Personal Protective Equipment (General Requirements and Eye, Face, Head, Foot, and Hand Protection)

Chapter 8 was written by Paige Laratonda.

INTRODUCTION AND SCOPE

Almost anywhere in general industry, you will see employees sporting some sort or array of personal protective equipment (PPE). PPE refers to anything an employee wears on his or her body to mitigate the effect of potential hazards that may reach them. OSHA sets requirements concerning the general use of PPE in the workplace, eye and face protection, head protection, foot protection, and hand protection. OSHA also regulates protective equipment related to personal fall arrest systems (fall protection) and equipment for electrical protection; however, these requirements are covered in different chapters of this book. Similarly, details about hearing protection devices (e.g., earplugs and muffs) are described in a different chapter.

PPE is common in all industries and can be lifesaving, but it is actually the last type of hazard control an employer should consider when weighing risk reduction options. Employers must first consider options to eliminate or replace a hazard. If these options are not feasible, the employer should consider ways to reduce risk with engineering or administrative controls. After these options have been applied or exhausted, only then should employers consider PPE. The "Hierarchy of Controls" was explained in the first chapter of this book and you learned that PPE was indeed the last line of defense from hazards. Let's consider the hazard of flying metal shavings and debris from the use of a lathe (a rotating machine that uses a cutting head to shape material). The employer should first consider if such a machine is absolutely necessary for operations. If it is, then the employer must first consider ways

125

to guard against flying debris such as enclosing the machine or shielding the point of operation. Even if a guard or shield can be used, there still may be the risk of flying debris being thrown at the operator who must stand close. So, the employer shall require the employee to also wear protective eyewear to reduce the risk of a foreign body from entering his or her eyes. Most commonly, PPE is layered with other controls.

Employers are required to assess the hazards of the workplace and determine if PPE is required for adequate protection from hazards. If required, employers must provide PPE to employees free of cost (with few exceptions). Eye and face protections (e.g., safety glasses, goggles, and face shields) are required when employees are exposed to eye or face hazards from flying particles, molten metal, liquid chemicals, acids or caustic liquids, chemical gases or vapors, or potentially injurious light radiation. Head protection (e.g., hard hats) is required when there is a potential for head injury from falling objects of electrical shock. Foot protection (e.g., safety shoes) is required where there is a danger of foot injuries due to falling or rolling objects, or objects piercing the sole, or an electrical hazard to the feet. Hand protection (chemical-resistant gloves, impact-resistant gloves, hot-work gloves, and leather-work gloves) is required when employees' hands are exposed to hazards such as skin absorption of harmful substances, severe cuts or lacerations, severe abrasions, punctures, chemical burns, thermal burns, and harmful temperature extremes.

PPE is usually layered with other controls and must be properly selected based on the hazards present, because PPE has limitations. The correct PPE must be selected based on the hazard evaluated. Then, procedures must be in place that require employees to wear PPE. Employees must actually use the PPE and wear equipment properly. Even then, PPE does not completely control the hazard, it just limits the severity of the hazard if it reaches the employee's body. PPE alone is typically not enough to reduce risk to an acceptable level, but it is often necessary to supplement hazard control efforts.

The OSHA standards covered in this chapter include:

- 29 CFR 1910 Subpart I
  - 29 CFR 1910.132—General requirements
  - 29 CFR 1910.133—Eye and face protection
  - 29 CFR 1910.135—Head protection
  - 29 CFR 1910.136—Foot protection
  - 29 CFR 1910.138—Hand protection
Standards not included

- Under 29 CFR 1910 Subpart I:
  - 29 CFR 1910.134—Respiratory protection
  - 29 CFR 1910.137—Electrical protective equipment
  - 29 CFR 1910.140—Personal fall arrest protection systems

NARRATIVE

Jeremy works in a machine shop and is a very talented machinist. He takes pride in his work. He has been welding, grinding, and working with machinery since he was a teenager. This occupation has provided him with a significant income. It has allowed Jeremy to purchase a beautiful home with a perfect overlook of his land. He even recently bought a nice sports car that he takes for rides on the weekend. Jeremy has a wife and two children that he looks forward to coming home to each day. The kids always come running to the front door when he returns from work, and Jeremy loves seeing the smiles on his kids' faces when he walks in.

Jeremy has been working as a machinist for twenty years and has a great safety record. His current supervisor does not strongly enforce the use of PPE, so Jeremy often finds himself cutting some corners when it comes to safety. Jeremy has not gotten hurt for twenty years, so why would he now or any other day? This seemed to be the motto Jeremy's supervisor lived by, so he started to adopt the same mantra. Supervisors' habits have a way of rubbing off on employees. Jeremy has worn prescription glasses since he was a kid. Every now and then, his supervisor will spend money on purchasing prescription safety glasses for Jeremy to use for work. These are specific glasses that are approved by OSHA, because they are rated for impact and have vision correction built into them. But it is currently the end of the year, and money is tight. At this time, Jeremy's prescription safety glasses are really scratched up from overuse, but his supervisor refuses to spend the money to replace them. So, Jeremy typically puts large, bulky safety glasses over his prescription eyewear. But today, Jeremy forgot these glasses at home. So, he decides he will be fine for one day wearing his normal non-protective prescription glasses. After all, he figures wearing prescription glasses must offer some protection and is better than nothing, but that is not the case.

Jeremy is cutting up some metal that he is preparing for a big project. The final step, at the end of the day, is grinding the metal to make sure the ends of his pieces are smooth. Jeremy tends to rush toward the end of the day, because he is so anxious to get home to his family. Jeremy approaches the grinding wheel. He knows in the back of his mind he should be wearing protective safety glasses and a face shield but wants to finish for the day. Besides, Jeremy thinks to himself, his supervisor hardly ever wears protective eyewear. And, Jeremy has never been struck by the metal in or near the eye before, so why would he today?

Sure enough, just a minute into the grinding task, the grinding wheel cracks and explodes, violently shooting shards of hard composite material into Jeremy's face. Because he is not wearing proper impact-rated eyewear, Jeremy's glasses shatter forcing hundreds of small shards of glass into both of his eyes. Jeremy screams in agony and quickly reaches to his back pocket for his phone to dial 911, but he cannot see his phone to dial the correct numbers. Luckily, his supervisor heard his screaming and comes rushing over to help dial for help. The ambulance came and takes Jeremy into the emergency room to perform surgery to extract the pieces of composite material and glass shards from his eyes. The surgeons know Jeremy's eyes were permanently damaged, and he would likely never see again.

Jeremy woke up in the recovery room to the sound of his wife's voice, but he could not see her. He could not see the tears streaming down her face, but from the fear in her voice, Jeremy knew it was bad. His wife tells him that they had to remove an eye and the other eye will be permanently blinded for the rest of his life. Jeremy's employer did not require employees to wear proper eye protection in the machine shop. Jeremy's supervisor did not set a good example. Jeremy was rushing at the end of his work shift, so he could be home to see the smiles on his children's faces. But instead, Jeremy will never drive his sports car again. Jeremy will never see his sons' smiling faces again. He never will see them play another ball game, and he will never see them grow into young men.

SUMMARY OF OSHA STANDARDS

General Requirements

PPE must be provided by the employer. This means an employer cannot require the use of safety glasses at work and then expect employees to buy their own. PPE must be provided by the employer to the employee free of cost. A few exceptions to this include: steel toe shoes, prescription, corrective vision eyewear (spectacles), everyday clothing, or items used to protect from weather (e.g., sunscreen, rain coats). Some employers choose to go above and beyond the requirements and pay for these items as well, but they aren't required to. If employees are required to wear PPE, but prefer to supply their own, they are allowed; however, employers must ensure the employees use it correctly.

PPE must be used properly and maintained in a sanitary and reliable condition. If employees do choose to bring their own PPE from home, it is still the responsibility of the employer to make sure it is adequate to protect them. For example, if an employee bought a stylish pair of blue safety glasses and brought them to work, the employer has the right and responsibility to forbid them if they do not meet OSHA requirements. Some employees who have to wear PPE every day do tend to want something more customized than what the employer purchases. This is allowed, as long as it still meets the

OSHA standards, employer requirements, and does not cause a hazard to the employee.

Employers are required to physically assess the workplace to determine if hazards are present, or are likely to be present, that may require the use of PPE. If they determine that PPE will be required, then the employer must identify the type of PPE needed, communicate the need to each affected employee, train each employee on the safe use of the PPE, and make sure that type of PPE is effective. The employer should have a physical document proving that they conducted the PPE hazard assessment. If you are ever unsure of what PPE should be worn, a good starting point is either a Safety Data Sheet of a chemical or the operator's manual of the piece of equipment an employee is working with.

A PPE hazard assessment is an essential and OSHA-required document in all workplaces where there are hazards present or likely to be present that warrant the need for PPE. Although there is no standard format, each employer must document (have in writing) a PPE hazard assessment. The following are required components of a PPE hazard assessment document:

- A statement that identifies the document as a PPE hazard assessment
- Locations in the facility or job classifications of employees that the PPE hazard assessment applies to
- Date of assessment
- The name of the person who conducted the assessment
- The hazard or exposures evaluated
- Details about what PPE is required in the location or for the job classification based on the hazard identified

Eye and Face Protection

The employer must ensure that each affected employee uses appropriate eye or face protection when exposed to eye or face hazards from flying particles, molten metal, liquid chemicals, acids or caustic liquids, chemical gases or vapors, or potentially injurious light radiation.

Each affected employee must use eye protection that provides side protection when there is a hazard from flying objects. Detachable side protectors (e.g., clip-on or slide-on side shields) meeting the requirements of this section are acceptable, as seen in figure 8.1. These are most commonly used to supplement prescription safety glasses, so that they meet this requirement.

Each affected employee who wears prescription lenses while engaged in operations that involve eye hazards must wear eye protection that incorporates the prescription in its design or wears eye protection that can be worn over the prescription lenses without disturbing the proper position of the prescription lenses or the protective lenses, as seen in figure 8.2.

Figure 8.1

Figure 8.2

Eye and face PPE shall be distinctly marked to facilitate identification of the manufacturer and must comply with the American National Standards Institute (ANSI) Z87 consensus standard. You should be able to physically locate a "Z87" marking somewhere on the eye/face wear. If you do not see the marking on the equipment, it is not properly rated and shall not be used for impact protection in the workplace. Figure 8.3 depicts the Z87 marking on shaded safety glasses rated for impact protection.

Beyond safety glasses, goggles can offer protection from some particular hazards such as chemical splashes and protection from hazardous gases and vapors. It is up to the employer to determine if tight-fitting goggles may be more appropriate based on adequate hazard assessment. Also, face protection may be required in addition to eye protection. The most common type of face protection is a face shield. A face shield must also be selected based on the hazards evaluated. For example, if a face shield is used for protection against impact, it must be ANSI Z87 rated. If a face shield is used for protection against chemical splash, it must be resistant to the chemicals in question. If a face shield is used for electrical protection, it must be rated for adequate voltage protection. Face protection is common in the practice of welding where

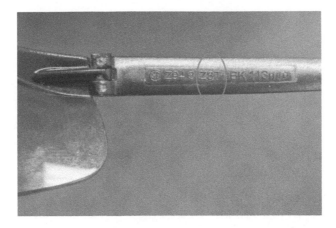

Figure 8.3

welding helmets must protect the face from molten metal and the eyes from
hazardous radiation. When a face shield is used for protection from flying
debris or splashes, eye protection is required underneath. When a face shield
is used for protection from flying debris or splashes, eye protection is required
underneath. Employers shall evaluate hazards present to determine the level
of eye and face protection required. Welding helmets and shaded visors are
covered in the welding and cutting chapter (chapter 16) of this book.

Head Protection

OSHA requires workers to wear hard hats when there is a potential for head
injury from falling objects, or electrical shock.

All head protection must meet the ANSI Z89.1 consensus standard. Hard
hats must have this rating, and it must be marked on the equipment as seen in
figure 8.4. ANSI Z89.1 defines two types of hard hats based on their method
of protection and establishes three classes of hard hats based on the level of
electrical hazard protection provided. The proper type and class of hard hat
shall be determined by employer hazard assessment.

Hard Hat Types: Types of hard hats are defined by the area of the head
that is protected.

- Type I offers protection to the top of the head
- Type II offers protection to the top and sides of the head (lateral impact)

Hard Hat Classes: The three classes are based on the level of protection they
provide from electrical hazards.

Figure 8.4

- Class G (General) hard hats are rated for protection up to 2,200 volts.
- Class E (Electrical) hard hats are rated for protection up to 20,000 volts.
- Class C (Conductive) hard hats do not offer electrical protection.

Each hard hat must have the following information clearly marked inside the hat:

- Manufacturer's name
- ANSI standard that the hard hat conforms with, such as "ANSI Z89.1-2009"
- ANSI type (type I or II) and class designation (G, E, or C)
- Size range for fitting
- Date of manufacture

While OSHA has no specific provision for an expiration date of a hard hat, manufacturers are allowed to determine if their equipment expires on a specific calendar date. In lieu of an expiration date, a generally accepted rule of thumb is to replace the hard hat suspension yearly and to replace the hard hat shell every five years. Harsh chemicals and extreme temperatures can make a hard hat degrade more quickly. Be sure to check with the manufacturer for guidelines on hard hat replacement and maintenance. Labels/stickers are acceptable to place on hard hats only if they do not adversely affect a hard hat's protective rating or make it more difficult to find potential defects and damage. For use of labels and stickers, refer to manufacturer instructions.

Foot/Leg Protection

Employees must use protective footwear when working in areas where there is a danger of foot injuries due to falling or rolling objects, or objects piercing the sole, or an electrical hazard, such as a static discharge or electric shock hazard. Again, employers are not required to provide (pay for) steel toe work shoes, but if an employee needs specific protective footwear for special hazards (e.g., for electrical hazard protection), then the employer is obligated to purchase those for them.

Protective footwear must comply with either ASTM International Standards F-2412 or ANSI Z41 consensus standards. These are usually seen on the inside of the tongue of the work boot.

Foot and leg protection choices include the following:

- Leggings—protect the lower legs and feet from heat hazards such as molten metal or welding sparks, safety snaps allow leggings to be removed quickly
- Metatarsal guards—protect the metatarsal area of the foot (top of foot) from impact and compression, made of aluminum, steel, fiber, or plastic, these guards may be strapped to the outside of shoes.
- Toe guards—fit over the toes of regular shoes to protect the toes from impact and compression hazards, they may be made of steel, aluminum, or plastic.
- Combination of foot and shin guards—protect the lower legs and feet and may be used in combination with toe guards when greater protection is needed.
- Safety shoes—have impact-resistant toes and heat-resistant soles that protect the feet against hot-work surfaces common in roofing, paving, and hot metal industries, the metal insoles of some safety shoes protect against puncture wounds.
- Electrically conductive shoes—provide protection against the buildup of static electricity. Employees working in explosive and hazardous locations such as explosives manufacturing facilities or grain elevators must wear conductive shoes to reduce the risk of static electricity buildup on the body that could produce a spark and cause an explosion or fire. Note: Employees exposed to electrical hazards must never wear conductive shoes.
- Electrical hazard, safety-toe shoes—are nonconductive and will prevent the wearers' feet from completing an electrical circuit to the ground.

Hand Protection

Employers shall select and require employees to use appropriate hand protection when employees' hands are exposed to hazards such as skin absorption of harmful substances, severe cuts or lacerations, severe abrasions, punctures, chemical burns, thermal burns, and harmful temperature extremes.

There are many types of gloves available today to protect against a wide variety of hazards. The nature of the hazard and the operation involved will affect the selection of gloves. The variety of potential occupational hand injuries makes selecting the right pair of gloves challenging. It is essential that employees use gloves specifically designed for the hazards and tasks found in their workplace, because gloves designed for one function may offer protection during a different task even though they may appear to be an appropriate protective device.

Sturdy gloves made from metal mesh, leather, or canvas provide protection against cuts and burns. Leather or canvas gloves also protect against sustained heat.

- Leather gloves protect against sparks, moderate heat, blows, chips, and rough objects.
- Aluminized gloves provide reflective and insulating protection against heat and require an insert made of synthetic materials to protect against heat and cold.
- Aramid fiber gloves protect against heat and cold, are cut- and abrasive-resistant, and wear well.
- Synthetic gloves of various materials offer protection against heat and cold, are cut- and abrasive-resistant, and may withstand some diluted acids. These materials do not stand up against alkalis and solvents.

Chemical-resistant gloves are made with different kinds of rubber or various kinds of plastic. A general rule: the thicker the glove material, the greater the chemical resistance, but thick gloves may impair grip and dexterity, having a negative impact on safety. Some examples of chemical-resistant gloves include:

- Butyl gloves
- Natural (latex) rubber gloves
- Neoprene gloves
- Nitrile gloves

It is very important that the right glove is selected based on the assessment of hazards (e.g., chemicals) in the workplace. It is also very important that gloves are only worn when it is appropriate. Sometimes gloves can pose more of a hazard if you aren't careful. For example, most rotating machinery can pull loose gloves or clothing into it and cause more harm than the gloves did good. In that case, you want to avoid wearing gloves. When in doubt, read the operator's manual, because it should indicate whether glove use is safe or not.

REVIEW QUESTIONS

1. Can an employee bring PPE from home? When are employers required to pay for PPE vs. When are employers not required to pay for PPE?
2. What line of defense is PPE? Should it be the first priority to protecting employees from hazards?
3. What are some limitations of PPE?
4. What document is required where PPE may be required in the workplace? What are the minimum elements of this document?
5. How can you determine if a hard hat is ready to be replaced?
6. When is eye protection required in the workplace? When is head protection required? When are gloves required?

REFERENCES

Occupational Safety & Health Administration [OSHA]. (2016). Regulations (Standards-29 CFR 1910.132). Retrieved from https://www.osha.gov/laws-regs/regulations/standardnumber/1910/1910.132

Occupational Safety & Health Administration [OSHA]. (2016). Regulations (Standards-29 CFR 1910.133). Retrieved from https://www.osha.gov/laws-regs/regulations/standardnumber/1910/1910.133

Occupational Safety & Health Administration [OSHA]. (2012). Regulations (Standards-29 CFR 1910.135). Retrieved from https://www.osha.gov/laws-regs/regulations/standardnumber/1910/1910.135

Occupational Safety & Health Administration [OSHA]. (2014). Regulations (Standards-29 CFR 1910.136). Retrieved from https://www.osha.gov/laws-regs/regulations/standardnumber/1910/1910.136

Occupational Safety & Health Administration [OSHA]. (1994). Regulations (Standards-29 CFR 1910.138). Retrieved from https://www.osha.gov/laws-regs/regulations/standardnumber/1910/1910.138

Chapter 9

Respiratory Protection

INTRODUCTION AND SCOPE

Respirators are a type of personal protective equipment (PPE) worn by an employee that protects him/her from a hazardous atmosphere. Because respirators are a type of PPE, they should only be used as a last line of defense for protection against hazardous atmospheres. This means respirators shall only be used when engineering controls (e.g., ventilation) or administrative controls (e.g., job rotation) are not enough to reduce the hazard exposure to safe levels or are not feasible. Respirators are also allowed to be used in emergency situations or in the interim when better controls are being implemented. There are several types of respirators that must be selected based on the evaluation of the potential hazards in the atmosphere.

Respirators must be appropriately selected based on the type and degree of atmospheric hazard. The potential atmospheric hazard must be evaluated to identify what the hazard is (e.g., oxygen deficient, presence of harmful contaminant). If a harmful air contaminant is present, its concentration must be measured or estimated to allow for adequate protection based on selecting the right respirator. Atmospheric hazard examples and air contaminant types include dust, fumes, gases, vapors, or oxygen-deficient atmospheres. The type of respirator needed depends on the type of hazard identified.

The primary types of respirators are air-purifying respirators and air-supplying respirators. Air-purifying respirators protect employees by purifying (filtering) the air before it is inhaled. Air-purifying respirators use filters, cartridges, or canisters to purify the air. Air-supplying respirators provide clean breathing air directly to the employee. Respirators can have tight-fitting facepieces, or respirators can be loose-fitting (e.g., hoods). It is imperative that the appropriate respirator is selected based on an evaluation of the hazards that might be present. Air-purifying respirators must be selected based on their ability to successfully filter out the air contaminants present. Not all air-purifying respirator cartridges, filters, or canisters are alike. Air-supplying respirators are typically used in high-hazard situations like emergencies or when the concentration of the air contaminant is unknown. Respirator protection factors (how effective they are) must be compared to exposure limits for the hazards present. If hazard evaluation reveals a respirator is required, the right respirator for adequate protection must be selected and provided to employees, and employers must implement a Respiratory Protection Program.

Any employees who must wear a respirator for protection must be included (follow the procedures of) a written Respiratory Protection Program. The program describes how the employer will comply with the OSHA Respiratory Protection Standard. The Respiratory Protection Program must establish procedures for the correct selection of respirators, medical evaluation and fit testing of users, training for respirator users, proper storage, maintenance, and use of respirators, and periodic evaluation of these procedures. Failure to establish and implement a Respiratory Protection Program and its required procedures can lead to employee illness or death. Employers are responsible for evaluating potential hazards for the need for respirators and ensuring employees follow the requirement of the program to prevent related incidents.

The OSHA standards covered in this chapter include:

• 29 CFR 1910.134—Respiratory protection

Not included:

• Under 29 CFR 1910.134
   ◦ Supplemental information for physicians and licensed health care professionals who perform or oversee medical evaluations for respirator users
   ◦ Qualitative and quantitative fit test procedures
   ◦ Requirements for respiratory protection procedures that apply specifically to interior structure firefighting
   ◦ Breathing air quality and use requirements for air-supplying respirators

NARRATIVE

Wendy has been the quality manager at a manufacturing plant for quite some time. Wendy started working in the plant as a quality control associate. Eventually, Wendy was named as a supervisor in her department. Wendy was very successful in her role as a supervisor and is now the quality manager, but Wendy is about to get many more responsibilities.

Wendy is a genuine, kind person who is always willing to pick up others' slack and be a team player. Recent budget cuts at the plant have led to some management-level employees being let go. To save money, top management names Wendy the new manager over quality, maintenance, and safety. Wendy is concerned with the workload and nervous about not being qualified to oversee these new areas. Despite her self-doubt and concern, Wendy commits to trying her best to do her part. She accepted the role and made a commitment to her coworkers. Wendy's number one concern is being a good boss for her employees, because she recognizes that they depend on her for successful employment. Wendy knows all of the employees that report to her. She is their friend. She knows their wives, husbands, kids, and life stories. Wendy is always thinking about others before herself and makes all of her decisions based on this perspective.

It is a Monday morning, and one of the maintenance supervisors tells Wendy he is going to have Tom put a coating on the walls of one of the pits under the production floor. The pit needs sealed, primed, and coated to effectively collect a by-product of the production process. Wendy is new to her role overseeing maintenance activities and safety programs. When it comes to maintenance operations, Wendy can rely on her maintenance supervisors' competence. When it comes to safety, the plant does not have many programs or procedures established, so Wendy has tried to get as much training as she can to help keep her employees safe. When Wendy hears about painting inside the pit, she knows enough to question the supervisor about the confined space and if the substance Tom will be using is dangerous. The supervisor reassures Wendy that they "got confined space entry procedures covered." The supervisor also admits that the chemical Tom will be using is "pretty nasty stuff." But, the supervisor helps Wendy feel better when he explains Tom will wear a respirator. The supervisor further assures Wendy that Tom has had training and has been fit tested to wear a respirator. The supervisor says, "they have plenty of respirators laying around," so Tom will be fine. Wendy is reluctant but does not have the time to further investigate the work or the maintenance department's work practices. The supervisor does a good job of smoothing-over Wendy, so she approves their work.

To complete the work, Tom grabs a respirator and starts applying a primer substance to the walls of the pit. The primer is pure methyl chloroform. Tom can notice that he smells the substance through his respirator, but it does not bother him much because it smells sweet. Methyl chloroform is toxic and can be deadly in confined spaces without proper respiratory protection. Tom was wearing an air-purifying respirator equipped to filter out solid, particulate dust. The respirator is virtually useless against the methyl chloroform vapors. Tom quickly becomes unconscious. Tom was being monitored by an attendant. This attendant jumps into the pit to rescue Tom, but he is not wearing any respiratory protection and is also overcome by dangerous vapors. Both employees die before an external rescue service can remove their bodies.

The employer failed to properly assess the hazards of the job. Tom was provided the wrong respirator based on the hazard present. Both employees did not have adequate training to know how to protect themselves from the hazard of the substance being used in the environment it was being used. Two employees lost their lives.

Wendy was busy in her office when the maintenance supervisor came to break the tragic news. As normal, Wendy greets the supervisor with a warm smile. The supervisor, a typically callous man, has tears in his eyes, and Wendy instantly feels a terrifying wave of grief. After Wendy hears the news, her heart is broken. Wendy knew these employees, she was their friend. She has been to their homes and played with their children. Wendy feels unimaginable guilt and blames herself for not further assessing the hazards of their job that day. Wendy becomes a shell of her former self. She resigns from her position at the plant and cuts ties with all her former employees and friends. While no one involved blames Wendy for what happened, Wendy blames herself and never forgives herself. Wendy is now unemployed, lives alone, and struggles with substance abuse. She feels she will never deserve another friend or relief from her depression because of what she failed to do.

SUMMARY OF OSHA STANDARD

Background Information

Types of Air Contaminants

Before understanding the OSHA Respiratory Protection Standard, it is helpful to know the different types of common air contaminants. Common types of air contaminants are:

- Air contaminants—harmful or unwanted airborne solids or liquids (e.g., harmful dusts, fumes, vapors)
- Particulates or dust—small airborne solids (e.g., aluminum metal, silica)
- Fumes—fine solids formed by condensation of vaporized metals (e.g., fumes of beryllium, copper, and lead)
- Smoke—carbon particles from incomplete combustion (e.g., smoke from burning wood)
- Mists—liquid droplets suspended in air (e.g., oil mist)
- Vapors—the gaseous phase of liquid which at normal temperature/pressure is liquid (e.g., gasoline vapors, alcohol vapors)

Types of Respirators and Similar Devices

In addition to understanding the types of air contaminants respirators may be used to protect against, it is helpful to be familiar with common types of respirators and similar devices. Respirators and similar devices can be grouped into three categories: non-respirators (similar devices), air-purifying respirators, and air-supplying respirators. Respirators can be further described as tight-fitting or loose-fitting. Tight-fitting respirators have a facepiece that fits, snugly on the user's face. Loose-fitting respirators are hoods or otherwise do not fit tightly on the user's face. Each type of respirator has advantages and disadvantages and particular applications.

Non-Respirators (Similar Devices)

First, there are two common devices or types of PPE that are not respirators but are worn over the nose or mouth. These devices are commonly mislabeled as respirators. Non-protection-rated nuisance dust masks and non-protection-rated surgical face masks are NOT respirators. Wearers of these devices do not have to be included in a Respiratory Protection Program.

Nuisance Dust Masks

A nuisance dust mask does not have a lab-tested (NIOSH-rated) filter, so there is no quantified level of protection provided. NIOSH stands for the National Institute for Occupational Safety and Health. NIOSH is the research group that sets procedures and test methods to evaluate respirators for their effectiveness. OSHA requires respirators used in the workplace to be rated and approved by NIOSH.

A nuisance dust mask is not rated by NIOSH and should not be confused with an OSHA-defined dust mask (filtering facepiece). A nuisance dust mask is not rated to provide any level of protection. However, an OSHA-defined dust mask (filtering facepiece) has a filter integral to the facepiece that is rated for protection to some degree. A nuisance dust mask is generally used to keep relatively harmless nuisance dust and allergens out of the nose and mouth. Wearing such a mask does not require employees to follow the procedures of a formal Respiratory Protection Program. Figure 9.1 depicts a non-rated, nuisance dust mask.

Surgical Face Coverings

Similar to nuisance dust masks, surgical face coverings do not provide any level of rated protection. Surgical face coverings are another example of PPE that are not respirators but are worn over the nose and mouth in some workplaces. Surgical face coverings usually do not have a lab-tested filter. In some specific uses, like in health care facilities, they can have a NIOSH-rated filter

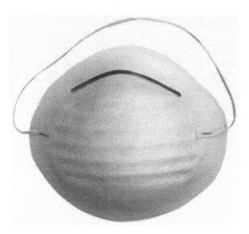

Figure 9.1

Figure 9.2

designed for protection. If that is the case, the device would be considered a respirator. But typically, these devices do not provide a rated level of protection and therefore are not a respirator. The surgical face coverings are designed to keep some airborne materials (not quantified) out of the nose and mouth and to contain the user exhales. Wearing such a surgical mask (no lab-tested filter) does not require employees to follow procedures of a formal Respiratory Protection Program. Figure 9.2 depicts a typical, pleated surgical face covering.

Air-Purifying Respirators

Air-purifying respirators use filters, cartridges, or canisters to remove contaminants from the air the user breaths in. Cartridges and canisters are attached to a respirator facepiece. They contain a filter or substance that

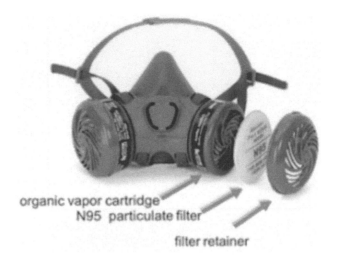

organic vapor cartridge
N95 particulate filter

filter retainer

Figure 9.3

removes particular contaminants from the air. Figure 9.3 shows an air-purifying respirator, with a dual vapor cartridge and particulate filter assembly. Air-purifying respirators are generally grouped into three types: filtering facepieces, air-purifying respirators (passive respirators), and powered air-purifying respirators (active respirators). If employees must wear any of these respirators for their protection, they must be included in and follow the procedures of a Respiratory Protection Program.

Filtering Facepiece Respirators (OSHA-Defined Dust Masks)

A filtering facepiece is indeed a respirator. OSHA refers to these as dust masks. OSHA-defined dust masks (filtering facepiece respirators) should not be confused with nuisance dust masks (not respirators). A filtering facepiece is designed for protection from certain air contaminants such as particulates and fumes. Filtering facepieces and particulate filters are assigned a letter rating of either N, R, or P and a numerical rating of either 95, 99, or 100. These ratings are combined to express their limitations (e.g., N95 and P100). These filter ratings must be certified by testing done by NIOSH. These ratings are described below:

- N = not resistant to oils
- R = somewhat resistant to oils
- P = strongly resistant to oils
- 95 = tested to filter out at least 95 percent of particles
- 99 = tested to filter out at least 99 percent of particles
- 100 = tested to filter out at least 99.97 percent of particles

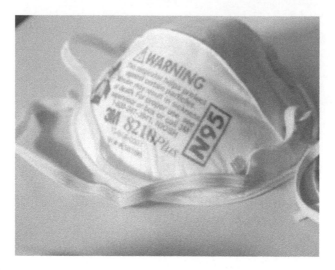

Figure 9.4

Figure 9.4 illustrates a filtering facepiece rated for protection (a respirator).

Employees who must wear a rated filtering facepiece (e.g., N95) must be included under and follow the requirements of a Respiratory Protection Program.

Air-Purifying Respirators (Passive Protection) and Powered Air-Purifying Respirators (Active Protection)

Air-purifying respirators (APRs) typically fit tight to the user's face. They usually cover half of the face (mouth and nose) or the full face. Typical APRs filter the air passively. This means when the user breaths in, the air is passed through a filter or filtering substance and purified before it reaches the user. Powered air-purifying respirators (PAPRs) can fit tightly on the user's face (nose and mouth only, or full face) or can be loose-fitting. PAPRs filter the air actively. PAPRs use a power source to suck in contaminated air, pass it through a filter or filtering substance to purify it, then provide fresh air to the user inside the facepiece or hood. Both APRs and PAPRs (whether tight- or loose-fitting) use filters, cartridges, or canisters to filter out air contaminants. The proper filtering device must be selected based on the air contaminant to be removed.

Figures 9.5 and 9.6 show air-purifying (tight-fitting) respirators. Figure 9.5 is a half-face. Figure 9.6 is a full-face respirator. Figure 9.7 shows two types of PAPRs (a half-face, tight-fitting respirator and a loose-fitting hood).

Employees who must wear APRs or PAPRs must be included under and follow the requirements of a Respiratory Protection Program.

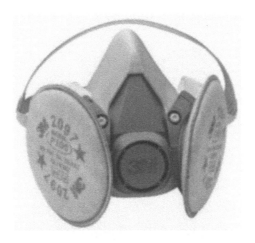

Figure 9.5

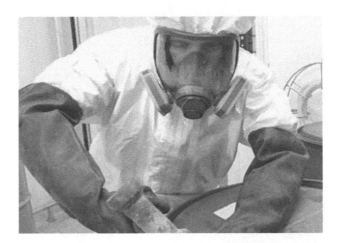

Figure 9.6

Air-Supplying Respirators (Airline Respirators and Self-Contained Breathing Apparatus Respirators)

Air-supplying respirators (ASRs) provide fresh breathing air directly to the employee. They do not use filters or filtering substances. Two types of ASRs are airline respirators and self-contained breathing apparatus (SCBA) respirators. ASRs are typically used in emergency situations (e.g., firefighting, rescue), where levels of air contaminants are unknown, or when the atmosphere is immediately dangerous to life or health (IDLH). The breathing air used with ASRs is typically stored in a container. If the container is separate from the employee and outside the hazardous environment and

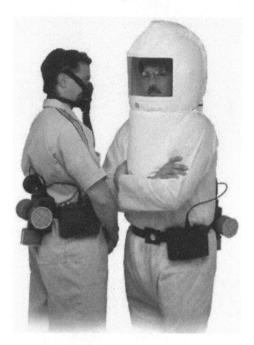

Figure 9.7

Figure 9.8

Figure 9.9

delivered by an airline (hose) to the facepiece, the respirator is referred to as an airline respirator. Figure 9.8 depicts an airline respirator. If the breathing air container is worn by the employee, it is an SCBA. Figure 9.9 depicts an SCBA. OSHA sets requirements for breathing air that is used with ASRs, but that information is omitted from this chapter. If your employer uses air-supplying respirators, you should be familiar with the requirements found in 29 CFR 1910.134. Simply speaking, the air supplied must be safe to breathe (e.g., necessary oxygen content). ASRs are most common in high-hazard situations. Employees who must wear ASRs must be included under and follow the requirements of a Respiratory Protection Program.

Allowable Respirator Use

When it comes to controlling the hazards of harmful dust, fumes, mists, gases, smokes, or vapors, the primary objective shall be to prevent the release and contamination of the atmosphere. This must be accomplished by engineering controls whenever feasible. Examples of engineering controls include: confinement of the operation that produces harmful substances, isolating workers from hazardous substances, general dilution ventilation, and local exhaust ventilation. When engineering controls are not feasible, while they are being implemented, or when administrative controls cannot reduce exposure to safe levels, respirators shall be used in compliance with the OSHA Respiratory Protection Standard. Generally speaking, respirators may only be used:

- If engineering/administrative controls are not enough to reduce atmospheric hazards to safe levels

- For temporary, non-routine tasks
- During emergency response
- In the interim while air contaminants are being sampled or measured
- In the interim while controls are being investigated or implemented

Respiratory Protection Program Overview

If employees are required to wear respirators, then the employer must implement a written respiratory program with the required procedures and elements
required for respirator use. The program must be administered by a designated and appropriately qualified program administrator. This person must
be explicitly designated within the written program. The written Respiratory
Protection Program shall include the following provisions as applicable to
the employer:

- Procedures for selecting respirators for use in the workplace;
- Medical evaluations of employees required to use respirators;
- Fit testing procedures for tight-fitting respirators;
- Procedures for proper use of respirators in routine and reasonably foresee-
  able emergency situations;
- Procedures and schedules for cleaning, disinfecting, storing, inspecting,
  repairing, discarding, and otherwise maintaining respirators;
- Procedures to ensure adequate air quality, quantity, and flow of breathing
  air for atmosphere-supplying respirators;
- Training of employees in the respiratory hazards to which they are poten-
  tially exposed during routine and emergency situations;
- Training of employees in the proper use of respirators, including putting on
  and removing them, any limitations on their use, and their maintenance; and
- Procedures for regularly evaluating the effectiveness of the program.

When included in the Respiratory Protection Program, employers must provide employees with respirators, training, and medical evaluation at no cost
to employees.

Program Exemption for Voluntary Use

There is one exception to employee inclusion under the Respiratory Protection
Program. This exception applies to employees who voluntarily use a respirator
or a filtering facepiece when it is not actually needed. Voluntary users do not
have to be included in the Respiratory Protection Program and follow all of the
provisions within. Voluntary use means the employee prefers to use a respirator
or filtering facepiece even though, through adequate assessment of the potential atmospheric hazards, there is not a hazard present that warrants the need

to wear one. This exception applies whether the respirator is a personal item brought from home or if it is available in the workplace. The following outlines minimum requirements for voluntary respirator users. Employers must:

- For respirators that are not filtering facepieces (e.g., APRs, PAPRs)
  - Provide users appendix D of the Respiratory Protection Standard ("Information for Employees Using Respirators When Not Required Under the Standard")
  - Employers must ensure the respirator itself does not create a hazard for the voluntary user via medical evaluation
  - Ensure the employee safely maintains, cleans, and stores the respirator in a way that does not create a health hazard for the user (in accordance with the manufacturer)
- For filtering facepieces (OSHA-defined dust mask e.g., N95)
  - Simply provide users appendix D of the Respiratory Protection Standard ("Information for Employees Using Respirators When Not Required Under the Standard") and have them follow the recommendation to ensure the facepiece does not create a hazard
- For nuisance dust mask (no rated filter for protection)
  - Nothing, employees can wear freely

If an employee voluntarily wears a respirator, and the employer assessment reveals one is not needed, then the employer is not financially responsible to provide the employee the respirator.

Atmospheric Hazard Assessment and Selection of Respirators under the Respiratory Protection Program

Employers are required to evaluate hazards and use-factors in the workplace to determine if a respirator is needed and to select the correct respirator. An evaluation of hazardous air contaminants must include a measurement or qualitative assessment of employee exposure to respiratory hazards and an identification of the chemical's state and physical form (type of air contaminant). Atmospheric hazards can be identified using a variety of direct-reading instruments that can detect and/or measure the concentration of air contaminants. To assess full-shift (eight hour) exposure air sampling is used. Air sampling consists of using an air pump to collect a sample of the air contaminant on a filter. This filter is analyzed to determine airborne concentration of contaminants, usually average exposure over the duration of a work shift. Figure 9.10 shows an employee arranged for air sampling. If the hazard cannot be identified, evaluated, or estimated (hazard unknown), then the atmosphere must be considered IDLH.

Figure 9.10

When evaluating use-factors, employers should consider the type of operation and environment the respirator will be used in, the level of activity required, and other external or personal factors. Employers shall only select and use NIOSH-certified respirators. The employer shall select, purchase, and stock respirators so they can provide the appropriate type, size, and fit for users. When needed, respirators must be provided to employees free of charge.

Selecting Respirators for IDLH Atmospheres

An atmosphere is IDLH if it contains levels or types of air contaminants air-purifying respirators cannot protect against. An atmosphere is IDLH if it is oxygen deficient or if levels of potentially hazardous air contaminants are unknown. In an atmosphere that is IDLH, a respirator must be provided that is a full-face pressure demand (admits air to facepiece by inhaling) SAR (airline or SCBA) certified by NIOSH with a minimum service life (air supply) of thirty minutes.

Selecting Respirators for Atmospheres That Are Not IDLH

When atmospheres are not IDLH but contain levels of air contaminants that can be potentially harmful, the correct respirator must be selected based on its ability to adequately protect from the type and concentration of the substance in the air. Employers must use the assigned protection factors (APFs) listed in table 1 of the OSHA Respiratory Protection Standard to select a respirator that meets or exceeds the required level of protection. This table of APFs is found in appendix A of this chapter.

When selecting a respirator, the employer must use APFs to ensure the Maximum Use Concentration (MUC) is not exceeded. An MUC is calculated by multiplying the permissible exposure limit of the air contaminant times the APF of the respirator being used. The MUC is then compared to the actual measured exposure concentration. The measured exposure concentration shall not exceed the MUC. Three examples of calculating MUC are provided below. Appendix A of this chapter is used (table of APFs) to calculate if protection is adequate:

- Example 1: Aluminum metal (total dust)
  - PEL: 15 mg/m^3
  - Respirator used: Half-face, tight-fitting, air-purifying respirator with P100 cartridges
  - APF:10
  - MUC: 15 mg/m^3 × 10 = 150 mg/m^3
  - Exposure Concentration: 20 mg/m^3
  - Result: The exposure concentration is less than the MUC. Protection is adequate.
- Example 2: Chlorine vapors
  - PEL: 0.5 ppm
  - Respirator used: Full-face, tight-fitting, air-purifying respirator with OV/Acid Gas cartridges
  - APF: 50
  - MUC: 0.5 ppm × 50 = 25 ppm
  - Exposure Concentration: 30 ppm
  - Result: The exposure concentration is GREATER than the MUC. NOT ADEQUATE—DANGER!
- Example 3: Hydrogen fluoride
  - PEL: 3 ppm
  - Respirator used: Loose-fitting, airline respirator
  - APF: 25
  - MUC: 3ppm × 25 = 75 ppm
  - Exposure Concentration: 50 ppm
  - Result: The exposure concentration is less than the MUC. Protection is adequate.

The employer must ensure the measured exposure concentration does not exceed the MUC. In addition to this comparison, the employer must ensure the respirator is appropriate for the chemical state and form of the contaminant. For example, for protection against gases and vapors, the employer must provide a respirator that is air-supplying or that is air-purifying and has an appropriate filter/filtering substance to adequately remove the contaminants present. Air-purifying respirators or their cartridges/canisters must

Figure 9.11

have an end-of-service-life indicator (ESLI). This is an indication of when the respirator or its filtering device must be replaced. Figure 9.11 shows an ESLI example. Not all ESLIs look alike, because manufacturers design them differently.

Not all APR cartridges, filters, and canisters are alike. The type of contaminants these devices can filter out depend on the filter or filtering agent inside. Cartridges and canisters are color-coded based on the contaminants they can filter. For example, a yellow color indicates protection against acid gases and organic vapors. A blue cartridge protects against carbon monoxide. A purple cartridge protects against particulates. Appendix B of this chapter provides an OSHA color-coding guide for the selection of cartridges based on the contaminant to be filtered out. Employers should also refer to manufacturer guidance on the specific cartridges used in the workplace to ensure proper selection of respirators and their filtering components.

For protection against particulates, the employer must provide a respirator that is air-supplying or that is air-purifying and has an appropriate particulate filter. For particulate protection, APRs (the filters/canisters equipped) must be equipped with a high efficiency particulate air (HEPA) filter. This is a filter that has been tested by NIOSH to filter at least 99.97 percent of particles.

Medical Evaluation for Employees in the Respiratory Protection Program

Respirators can make it harder to breathe and be physically taxing to wear. These effects can be compounded by the work environment (e.g., heat) and physical demands of work tasks. Respirator users must be medically evaluated to ensure wearing a respirator does not create a health hazard for them.

The medical evaluation must be performed or overseen by a physician or licensed health care professional (PLHCP). The medical evaluation shall at a minimum contain the information requested by the medical questionnaire found in 29 CFR 1910.134 appendix C. The purpose of the questionnaire is for a doctor to assess the health of an employee, provide further medical examination if needed, determine if an employee is fit to wear a respirator, or prohibit users from wearing a respirator. It is up to the medical evaluator to determine if further examination is needed beyond the standard questionnaire. OSHA provides supplemental information for the PLHCP performing medical evaluations for employees who wear a respirator, but this information is omitted from this chapter.

The employer shall obtain a written recommendation regarding the employee's ability to use the respirator from the PLHCP. The recommendation shall include the following information:

- Any limitations on respirator use related to the medical condition of the employee, or relating to the workplace conditions in which the respirator will be used, including whether or not the employee is medically able to use the respirator
- The need, if any, for follow-up medical evaluations
- A statement that the PLHCP has provided the employee with a copy of the PLHCP's written recommendation

After initial medical evaluation and determination of fitness to wear a respirator, additional medical evaluation is required if:

- An employee reports medical signs or symptoms that are related to the ability to use a respirator;
- A PLHCP, supervisor, or the respirator program administrator informs the employer that an employee needs to be reevaluated;
- Information from the Respiratory Protection Program, including observations made during fit testing and program evaluation, indicates a need for employee reevaluation; or
- A change occurs in workplace conditions (e.g., physical work effort, protective clothing, and temperature) that may result in a substantial increase in the physiological burden placed on an employee.

Respirator Fit Testing for Employees in the Respiratory Protection Program

If an employee must wear a tight-fitting respirator, they must be fit tested with the same make, model, style, and size of respirator they will use. This is typically done during the respirator medical evaluation with the medical service provider, but employers are allowed to conduct fit tests if they have individuals qualified to do so. Fit tests are required annually. So generally, the fit test and medical evaluation happen annually together. Qualitative or quantitative fit tests are allowed. OSHA provides mandatory qualitative and quantitative fit test procedures and guidance for how these tests must be conducted, but this information is omitted from this chapter. In summary, during a qualitative fit test, the employee wears the respirator and a harmless substance is sprayed into the air. The employee then says if he or she can smell or sense the substance through the respirator while performing certain tasks. During a quantitative fit test, a machine compares the environment outside the respirator to the environment within the respirator. The machine quantifies how well the fit is (if air is leaking in the facepiece past the seal). Fit testing shall be documented. The fit testing documentation shall include: the name of the employee tested, the type of fit test performed, the respirator used for testing, the date of the test, and the result of the fit test (pass-fail for qualitative tests and fit factor for quantitative tests).

Use of Respirators under the Respiratory Protection Program

OSHA sets requirements for prohibiting seal leakage of respirators, preventing removal of respirators in hazardous locations, ensuring effective respirator operation, and establishing procedures for use of respirators in actual or potential IDLH atmospheres.

For facepiece seal protection, employees with facial hair or facial deformities that might compromise the seal are not allowed to wear tight-fitting respirators. Anything that might interfere a respirator's function (e.g., block the exhalation or inhalation valve) is not allowed. Any PPE (e.g., safety glasses) shall not interfere with a respirator seal. All tight-fitting respirators must be subject to a user seal check each time they are put on. A seal check is done by covering the inhalation valve while breathing in to ensure the respirator sucks tightly into the face. Then, the employee covers the exhalation valve while exhaling to ensure the respirator is pushed away from the face from built-up positive pressure inside.

In addition to ensuring a proper seal for tight-fitting respirators, employers shall ensure all respirators remain effective for the duration they are worn. The employer shall ensure work conditions do not change in a way that might compromise the effectiveness of the respirator (e.g., additional air contaminant introduced that would exceed MUC). Employees must leave the area where a respirator is required if they:

- Must wash their faces or respirator facepiece, or otherwise remove their respirator

- Detect air contaminants entering their respirator
- Detect a leakage in respirator facepiece
- Need to replace a respirator filter, cartridge, or canister

When using respirators in IDLH atmospheres, the employer must ensure continued safe use. In IDLH atmospheres, the employer must:

- Have one or more employees located outside the IDLH atmosphere who ensure communication with those inside of it (e.g., an IDLH attendant)
- Ensure the IDLH attendant is equipped to provide (trained and physically equipped) rescue
- Ensure the IDLH attendant is equipped with the proper respirator to enter the IDLH atmosphere (e.g., SCBA) and appropriate rescue equipment (e.g., retrieval line)

OSHA provides requirements for respiratory protection procedures that apply specifically to interior structure firefighting. These requirements are omitted from this chapter.

Maintenance and Care of Respirators under the Respiratory Protection Program

OSHA requires employers to establish procedures for the cleaning, disinfection, storage, inspection, and repair of respirators. The employer shall ensure each respirator provided and used is clean, sanitary, and in working order. Generally, this should be done by referring to the respirator manufacturer's instructions. Respirators must be disinfected based on the following intervals:

- As necessary to keep sanitary if used excessively (e.g., each shift, daily)
- Before being worn by different individuals, if issued to more than one employee
- After each use, if used for emergency situations
- After each use, if used for fit testing or training

In addition to sufficient cleaning, respirators must be stored properly. Respirators must be stored in a way that prevents damage, deformation, contamination, dust, sunlight exposure, exposure to extreme temperatures, excessive moisture, and exposure to damaging chemicals. Respirators used for emergencies must be kept accessible to the area where the emergency may occur, stored in areas marked (labeled) for such storage, and stored in accordance with the manufacturer's instructions.

Beyond proper storage, all respirators shall be inspected before each use and during cleaning. All respirators used in emergency situations shall be inspected monthly and in accordance with the manufacturer's instructions. Emergency respirators shall be tested for proper function before and after each use. Respirators must be inspected as needed to ensure safe use and conditions. All respirators inspections must include a check of function, tightness of connections, condition of various parts such as the facepiece, head straps, valves, cartridges, and so on, and a check of any elastomeric (plastic/stretchy) parts for deterioration.

SCBA respirators shall be inspected monthly, and their oxygen cylinders shall remain fully charged (within 90 percent of the manufacturer's recommended pressure level). For respirators used in emergency situations, inspections must occur monthly and be documented in writing with the date of inspection, name of the person who did the inspection, required remedial actions, and some way of identifying the respirator (e.g., serial number). The documentation of inspection must accompany (e.g., on tag, label) the respirator and be kept with it during storage.

Finally, OSHA provides guidance for making repairs to respirators; however, it is not common for employers to make repairs to respirators, so these requirements are omitted from this chapter. Those making repairs must be qualified to do so and should do so in accordance with the manufacturer.

Training and Information for Employees in the Respiratory Protection Program

Employers must provide annual training to employees who use respirators. Training must provide the necessary information to ensure each employee who wears a respirator can demonstrate knowledge of at least the following:

- Why the respirator is necessary and how improper fit, usage, or maintenance can compromise the protective effect of the respirator;
- What the limitations and capabilities of the respirator are;
- How to use the respirator effectively in emergency situations, including situations in which the respirator malfunctions;
- How to inspect, put on and remove, use, and check the seals of the respirator;
- What the procedures are for maintenance and storage of the respirator;
- How to recognize medical signs and symptoms that may limit or prevent the effective use of respirators; and
- The general requirements of the OSHA Respiratory Protection Standard.

Employees must be retrained on the abovementioned concepts if changes in the workplace warrant the need or inadequacies in the employee's knowledge or use of a respirator are observed.

Respiratory Protection Program Evaluation and Recordkeeping

The written Respiratory Protection Program must be reevaluated periodically for its effectiveness and accuracy of procedures in the workplace. This includes a review of the written program and evaluation of procedures executed by employees. If corrective action is required, the employer shall ensure corrections are implemented to ensure the program's effectiveness. Employee feedback shall be used to evaluate the program. Factors to consider during program evaluation include but are not limited to respirator fit, proper respirator selection, proper respirator use based on workplace conditions, and proper respirator maintenance.

Employee medical evaluations regarding respirators shall be retained for the duration of the employee's employment plus thirty years. Fit test documentation must be retained until the next fit test occurs (e.g., annually).

REVIEW QUESTIONS

1. Name and describe four different types of air contaminants.
2. Compare and contrast a filtering facepiece with a nuisance dust mask.
3. Under what circumstances are respirators allowed to be used?
4. What are the applied protection factors for a half-face APR, a full-face PAPR, a full-face airline respirator, and a full-face SCBA?
5. What is the MUC for a substance with a PEL of 5 ppm while using a full-face APR?
6. How must hazards be assessed and measured to determine if a respirator is needed?
7. Describe the required procedures under a Respiratory Protection Program.
8. Describe the medical evaluation process for respirator users.
9. Describe the fit test process for respirator users, both qualitative and quantitative.
10. What is required for employees who are not required to wear a respirator but voluntarily wear one? (both for filtering facepieces and for APRs).

REFERENCES

Occupational Safety & Health Administration [OSHA]. (2011). Regulations (Standards-29 CFR 1910.134). Retrieved from https://www.osha.gov/laws-regs/ regulations/standardnumber/1910/1910.134

https://www.osha.gov/dts/shib/respiratory_protection_bulletin_2011.html

Appendix A—Assigned Protection Factors (29 CFR 1910.143 Table 1)

Type of respirator[1], [2]	Quarter mask	Half mask	Full facepiece	Helmet/ hood	Loose-fitting facepiece
1. Air-Purifying Respirator	5	[3]10	50	....	....
2. Powered Air-Purifying Respirator (PAPR)	....	50	1,000	[4]25/1,000	25
3. Supplied-Air Respirator (SAR) or Airline Respirator					
• Demand mode	....	10	50	....	....
• Continuous flow mode	....	50	1,000	[4]25/1,000	25
• Pressure-demand or other positive-pressure mode	....	50	1,000		
4. Self-Contained Breathing Apparatus (SCBA)					
• Demand mode	....	10	50	50	....
• Pressure-demand or other positive-pressure mode (e.g., open/closed circuit)	....	....	10,000	10,000	

Notes:

[1]Employers may select respirators assigned for use in higher workplace concentrations of a hazardous substance for use at lower concentrations of that substance, or when required respirator use is independent of concentration.

[2]The assigned protection factors in Table 1 are only effective when the employer implements a continuing, effective respirator program as required by this section (29 CFR 1910.134), including training, fit testing, maintenance, and use requirements.

[3]This APF category includes filtering facepieces, and half masks with elastomeric facepieces.

[4]The employer must have evidence provided by the respirator manufacturer that testing of these respirators demonstrates performance at a level of protection of 1,000 or greater to receive an APF of 1,000. This level of performance can best be demonstrated by performing a WPF or SWPF study or equivalent testing. Absent such testing, all other PAPRs and SARs with helmets/hoods are to be treated as loose-fitting facepiece respirators, and receive an APF of 25.

[5]These APFs do not apply to respirators used solely for escape. For escape respirators used in association with specific substances covered by 29 CFR 1910 subpart Z, employers must refer to the appropriate substance-specific standards in that subpart. Escape respirators for other IDLH atmospheres are specified by 29 CFR 1910.134 (d)(2)(ii).

Figure 9.12

Appendix B—OSHA Cartridge Protection Guidance

Contaminant	Color Coding on Cartridge/Canister
Acid gases	White
Hydrocyanic acid gas	White with 1/2 inch green stripe completely around the canister near the bottom.
Chlorine gas	White with 1/2 inch yellow stripe completely around the canister near the bottom.
Organic vapors	Black
Ammonia gas	Green
Acid gases and ammonia gas	Green with 1/2 inch white stripe completely around the canister near the bottom.
Carbon monoxide	Blue
Acid gases & organic vapors	Yellow
Hydrocyanic acid gas and chloropicrin vapor	Yellow with 1/2 inch blue stripe completely around the canister near the bottom.
Acid gases, organic vapors, and ammonia gases	Brown
Radioactive materials, except tritium & noble gases	Purple (magenta)
Pesticides	Organic vapor canister plus a particulate filter
Multi-Contaminant and CBRN agent	Olive
Any particulates - P100	Purple
Any particulates - P95, P99, R95, R99, R100	Orange
Any particulates free of oil - N95, N99, or N100	Teal

Figure 9.13

Chapter 10

Permit-Required Confined Spaces

INTRODUCTION AND SCOPE

Sometimes employees have to enter confined spaces to do work. Confined spaces are spaces that are large enough to enter, have limited means of entering and exiting, and are not intended to be inside for a long time. Confined spaces can have hazards related to the atmosphere, the energy sources inside of them, and the work being performed. If employees were to get hurt or maybe pass out while inside a confined space, and there was no emergency plan, incidents where employees might normally be able to recover could be fatal. The nature of being restricted and having limited ways to escape makes confined spaces inherently dangerous. Matters are made worse if there is hazardous energy inside or hazards related to the work being performed.

Employers are responsible for assessing the workplace and identifying spaces that would meet the definition of a confined space. They must identify and evaluate hazards related to these spaces. If a confined space has hazards that are not isolated or adequately controlled, then the confined space is a permit-required confined space (PRCS). As the name implies, PRCS can only be entered when there is a documented permit to assess allowable entry conditions and authorize entry. These permits have to be authorized (signed) by a designated supervisor. PRCS entries also require an attendant to watch over entrants. Because air contaminants can be especially dangerous in confined spaces, the atmosphere of a confined space must be monitored to make sure potential hazards are accounted for and the air is safe to breathe. If employees do have to enter PRCS, then a permit must be filled out, hazards must be continuously monitored, and rescue procedures have to be in place. If employees enter PRCS, then the employer must have a written PRCS program that documents procedures to comply with the OSHA standard and keep people safe during entries.

One way or another, all confined spaces need to be assessed for hazards (including air monitoring if there is potential for a hazardous atmosphere) to determine if the space will require a permit for entry. Again, the space is a PRCS if not all hazards of the space are adequately controlled or the work itself inside the space presents hazards that cannot be adequately controlled. Some employers will conduct an initial assessment of all spaces in their facility and determine which spaces are confined spaces and which spaces are PRCS. But the preferred, more-safe approach would be to consider all confined spaces permit-required until proven otherwise. This forces confined spaces to be assessed by specific tasks or before each entry. Given this approach, if all hazards can be adequately controlled for the task at hand, then the PRCS can be temporarily reclassified as non-permit-required (and this must be documented). If the space is determined to remain permit-required (hazards present), the employer can implement a permit system for entry, or the employer may not allow employees to enter PRCS at all. Some employers treat all confined spaces as PRCS regardless of hazards present, but this approach is not detailed in this chapter. The OSHA standard for confined spaces can be hard to understand but is summarized in this chapter in a way that hopefully makes it easier to digest.

The OSHA standards covered in this chapter include:

• 29 CFR 1910.146—Permit-required confined spaces

NARRATIVE

Stephen is a solitary, introverted person. He is thirty-one-years old and lives alone with his dog, Willow. Stephen has a family that cares about him and friends he sees on occasion, but he is happiest when he is spending time at home with Willow. He spends his free time working outside on his property, taking care of his horses, mending fences, and enjoying the outdoors with Willow. Stephen and Willow do everything together. Willow is always by Stephen's side when he is working in the pasture, taking a horseback trail ride, or cuddling up on the couch to read a book. Willow gives Stephen unconditional love and happiness. She relies on Stephen, and Stephen relies on Willow. They make each other happy.

Stephen works at the organic waste recycling plant in the nearest town. Each morning, before Stephen leaves for work, he kneels at the front door, rubs Willow's ears, and tells her he will be back soon. Willow spends the day checking on the horses outside and passing the time until her owner returns. Each day, when Stephen returns, he can see Willow in the window looking out for his truck. Willow greets Stephen at the door with a wagging tail, love, and kisses. Stephen can see Willow's soul behind her eyes and is thankful to have her companionship every day when he comes home from work.

One day, Stephen leaves for work, says goodbye to Willow, and is assigned the task of flushing out an underground sewage drainage system. Stephen has the unfortunate task of entering the confined space. At first, Stephen puts up with the strong smell of rotten eggs in the tunnel and makes his way deeper into the confined space. After some time, Stephen notices he can no longer smell the odor. It does not take long for Stephen to become dizzy, feel his body go into shock, and become incapacitated within the small drainage tunnel. Deadly levels of hydrogen sulfide are present in the tunnel and were the cause of Stephen's death.

The plant does not have a formal confined space program, and this was the root cause of Stephen's death. Employees like Stephen do not receive training on the potential hazards of confined spaces at the facility or the materials within. The hazards of the confined space were not assessed before Stephen entered nor was there a permit system to authorize entry. Air monitoring was not conducted before Stephen entered the space, and there was no rescue plan in the event of an emergency.

The day Stephen died, his body was recovered at 6 pm. That night, Willow stood by the door waiting for Stephen to come home. Willow sat by the door well into the night with her stomach growling with hunger and heart aching with anticipation for Stephen to return. Willow never saw her beloved friend again. It was not until two days later Stephen's brother came to his house to rescue the dog. Willow was taken care of, but she became shy and reserved. The dog's new home did not have what Willow longed for. She spent most of her days cooped-up inside. Even in her new home, years later, Willow spends most of the day sitting by the door waiting on Stephen to come home.

SUMMARY OF OSHA STANDARD

Types of Confined Space Entries

A confined space is any space that

- is large enough an employee can fit their whole body inside,
- has limited means of entrance and egress, and
- is not designed for continuous occupancy.

Examples of confined spaces include tanks, vessels, silos, bins, hoppers, vaults, and pits. Figure 10.1 provides an example of a confined space.

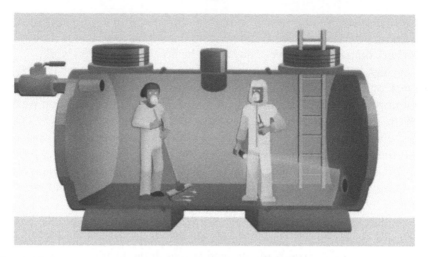

Figure 10.1

Figure 10.2

All employers shall evaluate the workplace to determine if any spaces are confined spaces (CS) or PRCS. A PRCS is a CS that has a hazard that is un-isolated or uncontrolled and can cause harm to entrants. Any space determined to be a PRCS shall be marked with the sign (or similar) seen in figure 10.2.

If an employer does not allow employees to enter PRCS (which is common, since they are so dangerous), then the employer must take measures to make sure employees do not enter them. If the employer does allow employees to enter PRCS, then they shall develop a written PRCS program that describes how the employer will meet the OSHA standard and keep employees safe during entry.

A PRCS is any CS that has an uncontrolled hazard. To enter a PRCS, entry has to be assessed, hazards identified and controlled, and entry authorized using a written permit. PRCS can be assessed and reclassified as a non-permit space if all hazards are isolated from the space and the assessment to justify the reclassification is documented. If the only uncontrolled hazard of a confined space is a potential unsafe atmosphere and that atmosphere can be maintained safe using continuous ventilation, then a permit is not required. It is important to note wearing a respirator does not adequately control or eliminate a hazard. When it comes to CS entry procedures, entries can be grouped into three categories. Table 10.1 describes the three categories of entries and summarizes what is required for each.

The data and hazard assessment process which is used to determine the type of entry must be made available to any CS entrants.

Any conditions that would make a PRCS unsafe to remove a cover or door (e.g., dangerous air contaminants) need to be eliminated before the cover or door is opened. When the cover or door is opened, measures need to be taken to prevent un-intended entry (e.g., temporary cover, guardrail to prevent falling in, and caution tape).

Air Monitoring

Before entering a confined space that has potential to have a hazardous atmosphere (or even sticking a head or hand in), the internal atmosphere of the space shall be tested with an air monitoring device that can immediately assess the hazards present in the air. This device shall be calibrated to work correctly in accordance with the manufacturer's directions. The air monitor device shall be able to assess the atmosphere for oxygen content, flammable gases and vapors, and potential toxic air contaminants. Typically, confined space air monitors test for, at minimum, carbon monoxide, hydrogen sulfide (a common toxic gas from decomposing organic material, that is, dead things), and some type of flammable substance. The air monitor must be able to test for the hazards that might be present. Different employers and facilities might need different air monitoring instruments. Figure 10.3 shows an example of a confined space air monitor. Non-PRCS spaces require, at minimum, initial monitoring to verify a hazardous atmosphere is not (unless it is objectively evident an atmospheric hazard is not possible and measurement is unnecessary). PRCS requires continuous, documented monitoring of the potential hazards. This means, if there is potential of a hazardous atmosphere, PRCS requires continuous air monitoring.

OSHA sets safe limits for oxygen content, flammable substances, and toxic air contaminants. Flammable gases, vapors, and so on are considered a hazard when they reach 10 percent of the lower flammable limit (LFL). An LFL is the minimum concentration or portion of air needed to be able to ignite. Oxygen content is determined to be hazardous if it makes up less than 19.5 percent of the air or makes up more than 23.5 percent of the air. Oxygen deficient atmosphere can cause asphyxiation (death from not taking enough oxygen into the blood). Oxygen enriched hazards pose an increased fire and explosion hazard from even a simple spark. Toxic air contaminants may not be present in concentrations

Table 10.1 Confined Space Entry Procedure Types

Entry Type:	Permit-Required	Forced-Air Ventilation	Non-Permit-Required
Description:	There is a hazard present that cannot be isolated or controlled	The only potential hazard present is quality of the air and that can be controlled with continuous, forced-air ventilation	There are no hazards or all hazards can be isolated or controlled such that they do not pose a risk
Requirements:	• Assessment of hazards to determine the type of entry needed • Written permit for entry • Continuous, documented air monitoring • Established roles (entrant(s), attendant, supervisor) • Training for those involved	• Assessment of hazards to determine the type of entry needed • Continuous, documented (if hazardous atmosphere is possible) • Attendant to monitor entrants • Training for those involved	• Assessment of hazards to determine the type of entry needed • Initial, documented air (if potential of hazardous atmosphere)

Figure 10.3

above their limits. OSHA sets ceiling limits, and/or short-term limits (fifteen-minute average exposure), and/or long-term limits (eight-hour average exposure) for certain air contaminants. It is up to the employer to determine what contaminants may be present and how to monitor those contaminants. OSHA sets limits for specific hazardous substances in 29 CFR 1910 Subpart Z.

Air monitoring must be conducted in a way that allows for complete and thorough measurement of the space. This might involve using an air monitor with a pump device that sucks air through a hose and delivers it to the measuring instrument. That way, air monitoring can be conducted without entry while employees remain safe. Figure 10.4 illustrates an employee conducting air monitoring. The hose of these monitors may need to be on rigid extension tools to reach in deep

Figure 10.4

enough and assess the area of the space furthest from the entrance. For vertical spaces, like tanks and manholes, gases can settle at different elevations based on their weight compared to air. For vertical spaces, employees must sample the bottom of the space, the middle of the space, and the top of the space to adequately measure the atmosphere and address this issue of gas distribution (stratification).

Specific Requirements for Permit-Required Confined Spaces

Entry Procedures and Permit System

Employers shall implement measures to prevent unauthorized entry into PRCS. Employers must identify and evaluate the hazards of a PRCS before employees enter them. There must be procedures in place that:

- Specify conditions for safe entry of each PRCS
- Isolate or control hazards within the PRCS
- Purge, flush, or ventilate any atmospheric hazards
- Prevent unauthorized or accidental entry into the PRCS
- Verify conditions are safe for the duration of the PRCS entry
- Provide equipment for safe entry into the PRCS

PRCS, as indicated in the name, requires a written permit for entry that is authorized (signed-off) by a supervisor. The permit must be completed before entry begins and must be made readily available near the portal of the entrance for the duration of the entry. The permit must authorize entry for only a specific time period.

Employers need to provide employees with equipment and training on equipment that is necessary to safely enter any PRCS they may get inside. This includes but is not limited to air monitoring equipment, ventilation equipment, communication devices, personal protective equipment (PPE), lighting equipment, barriers, equipment for safe entrance/egress (e.g., ladders), and rescue equipment.

PRCS entries and forced-air ventilation entries of CS require an attendant to be present. An attendant is an employee whose sole responsibility is to maintain safety during the entry. There must be formal, designated roles established for PRCS entry (entrants, attendants, and supervisor). A supervisor must authorize the permit before entry. The entry supervisor shall terminate entry and cancel the permit when entry operations are complete, or a condition (hazard) not allowed by the permit is present. A PRCS permit shall identify:

- The name of the permit space to be entered
- The purpose of entry

- The date/time/duration the permit is valid
- The names of authorized entrants
- The names of authorized attendants
- The name of the supervisor who authorizes entry
- The hazards of the space and how they will be isolated or controlled
- Acceptable entry conditions
- The results and times of air monitoring tests (initial and/or periodic)
- The rescue services available and how to summon rescue
- The communication procedures to be used amount entrants or between entrants and attendant(s) (e.g., radio, voice, and sight)
- Equipment required for safe entry
- Any other information necessary for safe entry

Appendix A provides an example of a blank entry permit.

Entry into PRCS (that has potential of a hazardous atmosphere) requires continuous air monitoring. This air monitoring needs to be documented periodically (e.g., every half-hour, every hour, every two hours) at an appropriate frequency based on potential hazards. Normally, this is documented on the permit for entry. Alternatively, some air monitors can record data automatically and be uploaded to a computer or online.

PRCS DUTIES AND RESPONSIBILITIES

Roles and responsibilities of entrants, attendants, and supervisors are outlined below. These responsibilities must be filled by employees serving in the established roles. They must be formally established and executed roles for entry into the PRCS. Employees who fill these roles must be trained on these responsibilities.

- Entrants:
  - Know the hazards of the PRCS they may enter
  - Know modes, signs, or symptoms that indicate exposure to a hazard
  - Properly use PRCS-related equipment
  - Communicate with the attendant to maintain safety
  - Alter the attendant whenever a hazard or condition is identified that is prohibited by the permit or a sign/symptom of hazard exposure is observed, an evacuation alarm or condition is observed
  - Evacuate when needed or directed by the attendant
- Attendants:
  - Know the hazards of the PRCS
  - Know the modes, signs, or symptoms that indicate entrant exposure to a hazard

- ○ Communicate with entrants as necessary to maintain safety
- ○ Monitor activity inside and outside space to identify any unsafe conditions (perhaps including air monitoring device)
- ○ Perform no other duties that will interfere with fulfilling the duty of an attendant
- ○ If the attendant detects a prohibited or unsafe conditions that warrant evacuation from the space, he/she shall direct entrant to immediately exit or summon rescue if needed
- ○ Warn and advise unauthorized entrants to stay away from the PRCS
- ○ Inform entrants if someone unauthorized enters the PRCS
- ○ Perform non-entry rescue if applicable (e.g., tripod and winch recovery for vertical spaces)
- • Supervisors:
  - ○ Know the hazards of the PRCS
  - ○ Know the modes, signs, or symptoms that indicate entrant exposure to a hazard
  - ○ Verifies entry permit conditions have been met and the space has been assessed of all hazards with appropriate hazard isolation and control measures taken
  - ○ Approves (signs-off/on) the entry permit based on satisfied conditions
  - ○ Terminates the entry permit if needed
  - ○ Verifies rescue procedures are in place and personnel are available
  - ○ Removes unauthorized personnel from the entry operation
  - ○ Determines if entry conditions and operations are in compliance with the permit

Rescue and Emergency Services

All PRCS require procedures to be in place that allow for the swift rescue of entrants if they are incapacitated. These rescue or emergency services can be conducted by the employer's employees or by an outside entity. For some PRCS, rescue services can be conducted by the attendant. For example, vertical PRCS (e.g., a manhole) rescue procedures may involve a tripod and winch. In this example, an entrant would wear a harness that is attached to a winching device and a tripod positioned over the space. If rescue is needed, the attendant can crank the winch and pull the entrant out. Figure 10.5 illustrates this type of rescue approach.

Regardless if the employer uses employees for rescue or if the employer uses an outside entity for rescue, the rescue service needs to be evaluated on its ability to perform its rescue and emergency services. The service needs to be evaluated on the ability to be summoned and performed duties in a timely manner. The evaluation needs to include the ability to perform rescue in the PRCS expected to be entered using the equipment expected to be used. The rescue

Figure 10.5

service must have the capability to rescue the victims in a time frame appropriate for potential hazards. Simply calling 911 may or may not be appropriate depending on the risk of hazards, ability of services to arrive in a timely manner based on those hazards, and if the rescue service has been evaluated on ability to perform rescue for the specific confined space being entered. Rescuers must have the equipment required to perform rescue. The rescue service shall be informed of the hazards they may encounter in PRCS. The rescue service must have access to assess any PRCS they may enter and be familiar with them, so they can develop appropriate rescue plans. If an employer designates their own employees for rescue, the employer shall take the following measures:

- Provide affected employees with the PPE needed to conduct permit space rescues safely and train affected employees so they are proficient in the use of that PPE, at no cost to those employees.
- Train affected employees to perform assigned rescue duties. The employer must ensure that such employees successfully complete the training required to establish proficiency as an authorized entrant.
- Train affected employees in basic first aid and cardiopulmonary resuscitation (CPR). The employer shall ensure at least one member of the rescue team or service holding current certification in first aid and CPR is available.
- Ensure affected employees practice making permit space rescues at least once every twelve months, by means of simulated rescue operations in which they remove dummies, manikins, or actual persons from the actual permit spaces or from representative permit spaces. Representative permit spaces shall, with respect to opening size, configuration, and accessibility, simulate the types of permit spaces from which rescue is to be performed.

Air Monitoring for Forced-Air Ventilation Entries

If a hazardous atmosphere can be controlled with forced-air ventilation (i.e., air supply and/or exhaust fan continuously running) and a hazardous atmosphere is the only hazard of a confined space, then a forced-air ventilation entry procedure can be used without a permit for entry. The forced air shall be directed to ventilate the space to remove any potential hazards in the atmosphere before entry. Care must be taken to prevent the forced air from simply circulating the hazard rather than removing it. This might involve using an air ventilation supply system coupled with an air exhaust system. The forced air shall be used for the duration of the entry, and the air shall be continuously monitored to ensure the ventilation is keeping the atmosphere safe. This air monitoring needs to be documented periodically (e.g., every half hour, every hour, or every two hours) at an appropriate frequency based on potential hazards. If a hazardous atmosphere is detected during entry, each employee shall exit the space immediately. Forced-air ventilation entries do not require a permit, but they require continuous air monitoring and an attendant. One specific caveat to this alternate entry procedure: When welding in a confined space, there shall be means provided for quickly removing the entrant in case of emergency (i.e., harness, lifeline, winch) and an attendant to perform this duty. Figure 10.6 depicts ventilation equipment used during a forced-air entry.

Figure 10.6

Reclassification of a Permit-Required Confined Space to a Non-Permit Confined Space

If a PRCS poses no actual or potential hazards (e.g., hazards eliminated or controlled), then the CS can be reclassified as a non-permit CS. This is common when employers consider all CS as PRCS until assessed and reclassified as non-permit spaces. If entry must be made into the space, before hazards are controlled, then PRCS requirements apply until reclassified as a non-permit space. For reclassification, the employer shall document the bases for making the determination. This document must include the date, location of space, and signature of the person making the determination. Typically, this form is referred to as a confined space hazard assessment document. This reclassification process is often tied to the control of hazardous energy (lockout/tagout). Often, the hazards in a confined space are related to energy sources. To isolate hazards within the place, energy needs to be isolated and lockout/tagout procedures must be applied. Once all energy is verified to be isolated, the PRCS can be reclassified as a non-permit space.

Training

PRCS training is required for any PRCS entrants, attendants, and supervisors. Training shall be provided to each employee before they are assigned PRCS-related duties, before they enter a PRCS, and whenever there is a change in the PRCS program. The training shall inform employees about the requirements of the PRCS OSHA standard, the requirements of the employer's PRCS program, and the duties of the roles employees will fill for PRCS entry. Employees shall have the right to inform of procedures that improve safety within the PRCS program.

Contractors

If a contractor must enter a PRCS of a host employer, then the host employer shall:

- Inform the contractor the space is a PRCS and that entry requires compliance with a PRCS program (either the contractor's program or the host's program)
- Inform the contractor of the hazards within the space or why the CS is permit-required
- Have a debrief with the contractor after entry to talk about any hazards encountered
- Provide any information to the contractor that is required for safe entry

REVIEW QUESTIONS

1. What is a confined space? Give three examples.
2. What is the difference between a confined space and a permit-required confined space?
3. Describe what confined spaces require initial air monitoring and which confined spaces require continuous air monitoring. What type of confined space entries require an attendant?
4. What three hazards must be sampled for when doing confined space air monitoring?
5. What are the main elements of a confined space entry permit?
6. What are the three roles of PRCS entry and what are the responsibilities of those roles?
7. Describe the process for reclassifying PRCS to be non-permit-required.
8. How do rescue/emergency service providers for PRCS need to be evaluated?

REFERENCES

Occupational Safety & Health Administration [OSHA]. (2011). Regulations (Standards-29 CFR 1910.146). Retrieved from https://www.osha.gov/laws-regs/regulations/standardnumber/1910/1910.146

Occupational Safety & Health Administration [OSHA]. (1993). Regulations (Standards-29 CFR 1910.146 Appendix D). Retrieved from https://www.osha.gov/laws-regs/regulations/standardnumber/1910/1910.146AppD

APPENDIX A—EXAMPLE PRCS
PERMIT (OSHA APPENDIX D-2)

```
                          ENTRY PERMIT

PERMIT VALID FOR 8 HOURS ONLY.  ALL COPIES OF PERMIT WILL REMAIN AT
JOB SITE UNTIL JOB IS COMPLETED


DATE: - -  SITE LOCATION and DESCRIPTION _____
PURPOSE OF ENTRY _____
SUPERVISOR(S) in charge of crews   Type of Crew Phone #
_____
_____

COMMUNICATION PROCEDURES _____
RESCUE PROCEDURES (PHONE NUMBERS AT BOTTOM) _____
_____

* BOLD DENOTES MINIMUM REQUIREMENTS TO BE COMPLETED AND REVIEWED
PRIOR TO ENTRY*


REQUIREMENTS COMPLETED                        DATE          TIME
Lock Out/De-energize/Try-out                  _____         _____
Line(s) Broken-Capped-Blanked                 _____         _____
Purge-Flush and Vent                          _____         _____
Ventilation                                   _____         _____
Secure Area (Post and Flag)                   _____         _____
Breathing Apparatus                           _____         _____
Resuscitator - Inhalator                      _____         _____
Standby Safety Personnel                      _____         _____
Full Body Harness w/"D" ring                  _____         _____
Emergency Escape Retrieval Equip              _____         _____
Lifelines                                     _____         _____
Fire Extinguishers                            _____         _____
Lighting (Explosive Proof)                    _____         _____
Protective Clothing                           _____         _____
Respirator(s) (Air Purifying)                 _____         _____
Burning and Welding Permit                    _____         _____
Note:  Items that do not apply enter N/A in the blank.
```

```
                **RECORD CONTINUOUS MONITORING RESULTS EVERY 2 HOURS
CONTINUOUS MONITORING**    Permissible    _____
TEST(S) TO BE TAKEN        Entry Level
PERCENT OF OXYGEN          19.5% to 23.5% ___ ___ ___ ___ ___ ___ ___ ___
LOWER FLAMMABLE LIMIT      Under 10%      ___ ___ ___ ___ ___ ___ ___ ___
CARBON MONOXIDE            +35 PPM        ___ ___ ___ ___ ___ ___ ___ ___
Aromatic Hydrocarbon       + 1 PPM * 5PPM ___ ___ ___ ___ ___ ___ ___ ___
Hydrogen Cyanide           (Skin)  * 4PPM ___ ___ ___ ___ ___ ___ ___ ___
Hydrogen Sulfide           +10 PPM *15PPM ___ ___ ___ ___ ___ ___ ___ ___
Sulfur Dioxide             + 2 PPM * 5PPM ___ ___ ___ ___ ___ ___ ___ ___
Ammonia                           *35PPM  ___ ___ ___ ___ ___ ___ ___ ___
```

* Short-term exposure limit: Employee can work in the area up to 15 minutes.

+ 8 hr. Time Weighted Avg.: Employee can work in area 8 hrs (longer with appropriate respiratory protection).

REMARKS:_____

```
GAS TESTER NAME        INSTRUMENT(S)        MODEL           SERIAL &/OR
  & CHECK #              USED             &/OR TYPE          UNIT #

_____      _____    _____      _____

_____      _____    _____      _____
```

```
        SAFETY STANDBY PERSON IS REQUIRED FOR ALL CONFINED SPACE WORK
SAFETY STANDBY    CHECK #   CONFINED                 CONFINED
  PERSON(S)                  SPACE      CHECK #        SPACE      CHECK #
                           ENTRANT(S)                ENTRANT(S)

_____  _____  _____  _____   _____   _____

_____  _____  _____  _____   _____   _____
```

SUPERVISOR AUTHORIZING - ALL CONDITIONS SATISFIED_____
                        DEPARTMENT/PHONE _____

AMBULANCE 2800 FIRE 2900   Safety 4901 Gas Coordinator 4529/5387

Chapter 11

The Control of Hazardous Energy (Lockout Tagout)

INTRODUCTION AND SCOPE

The control of hazardous energy is essential to make sure employees who are working on equipment do not become accidentally hurt while doing their job. Hazardous energy can take many forms like moving parts, high temperatures, chemicals, electricity, falling objects, and so on. Energy is anything that can be a hazard, fuel a hazard, or activate a hazard. All of these sources of energy must be adequately controlled to prevent accidental release or activation while people are working on machines and equipment.

The control of hazardous energy is achieved through an energy control program and energy control procedures. Typically, energy sources must be physically locked, so they can't be accidentally turned on while people are in the danger zone of a machine doing their job. Tags accompany locks, so there is individual accountability for who applied the lock, and so that person can be identified. Only the person who applied a lock may have the key to unlock it. This method of individual protection prevents someone else from activating the hazard while the person is still working on the equipment. The use of locks and tags to control, isolate, or "lockout" hazardous energy sources is called "Lockout Tagout."

The OSHA standards covered in this chapter include:

- 29 CFR 1910.147—The control of hazardous energy (lockout/tagout)

NARRATIVE

Cody is a multi-talented young man. If he's not at work turning wrenches, he is under the hood with his dad working on his car. With the little free time he has, Cody makes his fingers bleed practicing guitar. By the age of twenty-one, he has earned a lead-mechanic position at the company he works for. He has a wife and two daughters that he provides for. He exercises his passion for music every weekend by playing guitar for his band on local stages and in local bars. He does what he loves and provides for his family with his two hands. But in a gruesome incident, Cody lost more than just his callous hands.

It's Monday morning. At 5 am, Cody's alarm goes off, and he quiets it quickly to not disturb his wife. He kisses his sleeping wife and two daughters goodbye, and he drives to work. Cody clocks in. The maintenance supervisor explains that an operator, once again, jammed the shredding machine. The shredding machine uses rotating blades to chop material into smaller bits. It is a job Cody has done 100 times. So, Cody turns off the machine, gathers the tools he needs, and removes the guards on the machine to adjust the blades and remove the jam.

While Cody begins the routine maintenance job, a new employee starts his first day. The new employee who works in the scrap department is assigned to the shredding machine. The operations supervisor tells the new employee: "Run the shredder. You can't mess that up. Turn the machine on. Put material in the feeder. Press the "START" button. Repeat." The new employee heads to the shredding machine operating station.

The company Cody works at does not have a formal safety program. It does not provide employees with safety training. Employees are expected to figure it out as they go and do their job. At the shredder station, the new operator turns the power on to the shredder. The new employee cannot see Cody working on the other side and has no indication Cody is working on the blades. Cody has no idea the machine was powered on, until it's too late. Following his supervisor's directions, the new employee loads material into the feeder and presses the START button. In an instant, while Cody is removing the jam and making the adjustment, the razor sharp blades of the shredder are unexpectedly energized. The blades move viciously, and in the blink of an eye, Cody's hands and forearms are shredded by the unforgiving, cold steel.

Both of Cody's arms had to be amputated just above the elbows. The failure to control the hazardous energy and prevent the unexpected start-up of dangerous equipment while Cody performed his maintenance task resulted in catastrophe. Cody was permanently disabled. He lost the ability to provide for his family in the way he felt he was born to. Cody embraces the moments his daughters hug his waist but would give anything to fully embrace around them. Cody is now forced to look up at his replacement guitarist on the weekends while sitting in the audience, and when he does, his only thought is that he wishes he could clap for his friends on stage.

SUMMARY OF OSHA STANDARD

Application

The OSHA standard for the control of hazardous energy applies to the service and maintenance of machines and equipment when unexpected activation of the machine, equipment, or energy could harm someone performing a task. The standard only applies to service and maintenance tasks. Examples of service and maintenance tasks include lubrication, cleaning, removing jams, and anything that's not part of normal, routine, and repetitive operations. Any time someone needs to remove a guard revealing a hazard and/or place his/her body in a danger zone, then any energy that could potentially cause harm to the employee needs to be adequately controlled. Adequate control is typically achieved by locking energy in a safe state and applying tags to show who applied the locking mechanisms. Figure 11.1 shows a typical application of lockout/tagout (LOTO).

Two common exceptions to LOTO exist. The standard does not apply to tasks part of normal production and operations that happen all the time. However, for this exception, there needs to be controls in place in some way

Figure 11.1

that prevent injury. The employee can't just "quickly stick their hand in the danger zone before it's too late." For example, let's say that after every stroke of the machine, an employee needs to make a minor adjustment in the danger zone of the machine. When that adjustment is made, there need to be safeguards that prevent injury, like a sensor that prevents the machine from moving when his/her hands are in the point of operation. The standard also does not apply to work on equipment that can be simply unplugged to isolate all hazards and energy involved. However, for this exception, there needs to be only one person performing the work who can make sure the equipment is not plugged in by someone else. For example, say an employee needs to change the blade on a table saw that receives all of its energy from a plug-in cord. If that employee is the only person around, he/she can just unplug the machine, change the blade, and then plug it back in. For these so-called "exemptions," perhaps a better term is "alternative method of the control energy." For these alternative methods, you still need to control the hazard—you just do not have to apply physical locks and tags. In almost all other scenarios of service and maintenance tasks, locks and tags need to be affixed to isolation devices to prevent energy from harming the employee doing the job. An isolation device is a thing that physically prevents the energy from entering the danger one or affecting the employee. Two examples of isolation devices include: electrical switches that interrupt the flow of electricity and valves that stop the flow of chemicals, liquids, or air. Figure 11.2 is an electrical disconnect (knife switch) and figure 11.3 is a ball valve.

Figure 11.2

Figure 11.3

Both are isolation devices because they physically separate the source of energy from getting to the equipment.

Energy Control Program and Procedures

OSHA requires employers to have an energy control program when employees perform service or maintenance on machines and equipment that have hazardous energy sources. That means there must be evidence there are procedures in place to control energy during maintenance and service activities, the company must provide employee training on energy control, and there must be proof they are doing things to prevent the accidental start-up of equipment that can harm employees while they do their job.

In addition to having hazardous energy, some machines have stored energy after shutdown, and some have more than one energy source. Stored energy is energy that has the potential to cause harm even after the energy source is shut off. For example, you might shut off the air that leads to a machine, but there still can be air in the pipes that could be released and cause harm to someone doing maintenance work. Some equipment can be very large and complex with multiple energy sources like hydraulic pressure, air, electrical energy that moves things, and gas or chemicals that serve some function. Any time a machine, equipment, or system has either the potential of stored energy or has more than one energy source then there must be a documented energy control procedure (ECP). An ECP describes how the machine is shut down and how the hazardous energy is adequately controlled or isolated. OSHA gives companies guidance in 29 CFR 1910.147 Appendix A on typical minimal lockout procedures they can include in ECPs.

Related to LOTO and ECPs, there are three pieces of hardware that should be defined: isolation devices, lockout devices, and tags. Isolation devices are things that physically break the path of energy like electrical switches and pipe valves. Examples are seen in figures 11.2 and 11.3. Lockout devices are things that hold the isolation device in a safe position. Typically a lockout device is a lock maybe used in combination with some other hardware to assist in keeping the isolation device in the off (or safe) position. Examples of these lockout devices are seen in figures 11.4 and 11.5. Tags typically accompany the lockout device applied to the isolation device. Tags should include the name of the person who applied it (person in potential danger if machine was energized), and tags must say something similar to "Do Not Operate." Both lockout devices and tags need to be durable for the environment they're used in. They also need to be identifiable and standardized. This means that when an employee sees these locks and tags, they have to know that the only purpose of using them is for energy control and LOTO. This is often done with color-coding. It should be noted that it is not recommended to use standard LOTO equipment for purposes other than protection during service and maintenance tasks (e.g., using for security). Doing so degrades the meaning of LOTO devices.

Employers are required to do periodic, documented inspections of employees performing established ECPs and executing the control of hazardous energy. These inspections need to occur annually. Of course, any deviations or inadequacies identified need to be corrected. The company needs to be able to prove to OSHA that this occurs each year.

Training

Training is required for those who perform LOTO and work around LOTO operations. A LOTO authorized employee means one who is authorized

Figure 11.4

Figure 11.5

to conduct LOTO and perform the service or maintenance task. These employees must receive training on the purpose and function of the overall energy control program. Training shall give these employees the skills they need to execute LOTO and energy control. For authorized employees, the training shall generally include the recognition of hazardous energy, types and magnitudes of energy in the workplace, and methods of energy control or isolation. A LOTO-affected employee means an employee who may not perform LOTO but works around it and must know about the program. For affected employees, the training shall generally include the purpose and use of LOTO and ECPs, where ECPs might be used, instructions on the general procedure for controlling energy, and the prohibition of attempting to remove energy control devices and starting up the machine during a state of LOTO.

General LOTO Procedure

OSHA describes a general LOTO procedure. This includes the following mandatory sequence:

1. Notify affected employees the machine will be shut down and LOTO applied.
2. Prepare for shutdown by removing employees or tools and gathering equipment needed.
3. Shut down the equipment properly and safely.

4. Isolate any energy sources that can cause harm for the task that needs to be performed.
5. Apply lockout devices and tags on isolation devices to ensure the energy remains controlled during the job (isolated device is in off/safe position).
6. Relieve or release any stored or potential energy safely.
7. Verify energy is adequately controlled or isolated by trying to activate the energy sources while they are isolated.
8. Perform service or maintenance task.
9. Notify affected employees the machine will be re-energized and ensure all employees and tools are clear and it is safe to do so.
10. Remove LOTO devices and restore the machine to operating conditions safely.

Step 7 is especially important. Some companies refer to LOTO as Lock, Tag, and Verify. This is because isolation devices can fail. Sometimes it can appear that energy is turned off, but you cannot know for sure until you try to actually run the machine. It is required that de-energization is verified in some way (try to start machine), before performing the maintenance task.

Specific Topics

There are a few specific topics to cover when it comes to the control of hazardous energy. Whenever contractors or outside personnel come in to a company to work on their machines and equipment, the employers must trade procedures related to LOTO. The outsiders need to know the hazards and how to control them. Either employer's energy control program can be followed. Often, more than one person works together to perform the service or maintenance task. When this occurs, there must be procedures in place to ensure each and every person has individual protection. For example, each person must hold, on their person, a key that ensures the energy cannot be released unless they unlock the isolation device(s). One common way to do this is using a gang lockbox seen in figure 11.6. These lockboxes involve a designated primary responsible person. Typically, this primary responsible person locks out all isolation devices and puts the keys in a box. Then, all the employees who will perform the work place their own lock on the box to make sure the keys remain in the box until each and every person takes their lock off the box.

Lastly, there needs to be procedures in place to ensure energy remains controlled when there are shift changes. Typically, this is done by making

Figure 11.6

sure the second (incoming) shift first applies LOTO before the first (outgoing) shift removes LOTO.

REVIEW QUESTIONS

1. Give three examples of service or maintenance tasks that would require LOTO.
2. Give two examples of tasks that would require energy control (alternative safeguards), but not necessarily LOTO.
3. Give four examples of types or forms of energy. What is stored or potential energy?
4. When is an energy control procedure (ECP) required?
5. Describe the general sequence that OSHA prescribes for a LOTO procedure.
6. What is meant by "individual protection" must be maintained during LOTO?
7. What training is needed for someone who applies LOTO? What training is needed for someone who doesn't apply LOTO but works around machines that get locked out?

REFERENCE

Occupational Safety & Health Administration [OSHA]. (2011). Regulations (Standards-29 CFR 1910.147). Retrieved from https://www.osha.gov/laws-regs/regulations/standardnumber/1910/1910.147

Chapter 12

Fire Protection

INTRODUCTION AND SCOPE

Many industrial settings involve processes and equipment that use or generate heat, flames, sparks, or other ignition sources. These work environments can have fire hazards such as flammable liquids, gases, and vapors. Depending on the presence of certain fire hazards, employers must take measures to implement fire protection systems.

Some employers may choose to only allow trained individuals to fight fires or even provide fire rescue services. In this case, an employer may choose to organize a fire brigade as a fire protection system. This is a group of organized employees trained to fight and extinguish fires. Where employers organize an employee-fire brigade, the employer shall provide members of the brigade with adequate safety training, written safety procedures, and necessary protective equipment. Usually, depending on their job requirements, fire brigade members must be trained in rescue procedures and be included in a Respiratory Protection Program. If a fire brigade is established, it must be referenced in an employer's Emergency Action Plan and Fire Prevention Plan. Fire brigades are most common in heavy industrial plants with a high risk of fire hazards.

Portable fire extinguishers may be required in the workplace based on specific fire hazards and as required by local fire codes. Where the employer provides fire extinguishers for use by employees in the workplace, the employer shall ensure proper selection and placement of fire extinguishers. Fire extinguishers are classified based on the types of fires they are capable of extinguishing. Fire extinguishers must be readily accessible for use. Fire extinguishers must be inspected regularly and properly maintained. Employers must visually inspect fire extinguishers each month to ensure they are fully charged and ready for operation. Fire extinguishers must also receive annual maintenance by qualified technicians. Some extinguishers must be hydrostatically tested to ensure the integrity of the extinguisher under pressure. All inspections and maintenance of extinguishers must be documented and available upon request. Employees expected to use portable fire extinguishers must receive annual training about proper use. The employer may choose, as part of their Emergency Action Plan and Fire Prevention Plan, to not allow employees to fight fires with portable fire extinguishers and may direct them to immediately evacuate rather than attempt to extinguish fires.

Finally, sprinkler systems, fire detectors, and fire alarm systems may be required in certain workplaces based on the presence of certain hazards and as required by local fire and building codes. Sprinkler systems need to be tested regularly. Sprinkler heads must be protected from damage. There must be at least 18 inches between sprinklers and objects below or beside. Fire detectors and alarms must be kept in working order where installed. Manual fire alarms must be readily accessible and not obstructed.

For all fire protection equipment (extinguishing systems, sprinklers, fire detectors, and fire alarm system) there may be specific OSHA standards that require the use of such systems. More specifically, local fire and building codes will also dictate what workplaces will need to install such fire protection systems. Safety professionals should be familiar with codes in the local municipality or state in which they work.

The OSHA standards covered in this chapter include:

- 29 CFR 1910 Subpart L:
  - 29 CFR 1910.155—Scope, application, and definitions applicable to this subpart
  - 29 CFR 1910.156—Fire brigades
  - 29 CFR 1910.157—Portable fire extinguishers
  - 29 CFR 1910.159—Automatic sprinkler systems
  - 29 CFR 1910.164—Fire detection systems

The OSHA standards not included:

- Under 29 CFR 1910 Subpart L:
  - 29 CFR 1910.158—Standpipe and hose systems
  - 29 CFR 1910.160—Fixed extinguishing systems, general
  - 29 CFR 1910.161—Fixed extinguishing systems, dry chemical
  - 29 CFR 1910.162—Fixed extinguishing systems, gaseous agent
  - 29 CFR 1910.163—Fixed extinguishing systems, water spray, and foam
  - 29 CFR 1910.165—Employee alarm systems

NARRATIVE

Alex works at the front desk at a facility that does research on automated technology in vehicles (e.g., self-driving or autonomous systems). Alex is a single dad who raises his son alone. Alex's son is his whole world. On the wall behind Alex's desk, he hangs all the pictures his son draws for him.

When things are slow at the front desk, Alex helps the electronic engineers by doing some soldering to help build circuit boards. Soldering uses a hot iron to melt and fuse metal. Alex is working on the circuit boards when a researcher comes in and requests some help with a large shipping order. Alex sets his soldering iron down and leaves his desk to help the employee. Without realizing it, Alex sets his hot soldering iron on a pile of paperwork. It does not take long for the pile of paper to start smoldering and catch fire. The surrounding cardboard boxes near Alex's desk start to catch fire as well.

An employee runs to the shipping area and alerts Alex of the fire. Alex runs back to his desk, grabs the nearby fire extinguisher, and dowses the fire in the dry chemical powder. With some persistence, the fire goes out. Alex looks at his son's pictures hanging on the wall and notices the char of the flames is just below the delicate paper drawings. Alex is shaken by the event but breathes a deep sigh of relief knowing his most prized possessions were spared thanks to his fire extinguisher training.

SUMMARY OF OSHA STANDARDS

Fire Brigades

Some employers may choose to establish employee-led fire brigades to assist with extinguishment of fires. If so, establishment of the fire brigade must be covered in the employer's Emergency Action Plan and Fire Prevention Plan. A fire brigade is a private, industrial fire extinguishing service made up of an organized group of employees who are knowledgeable, skilled, and trained in basic or advanced firefighting operations. When a fire brigade is organized by an employer, the employer shall have a written policy that includes the following information:

- A statement establishing the existence of fire brigade
- The basic organizational structure
- The type, amount, and frequency of training required by brigade members
- The expected number of employees as part of the brigade
- The functions of the brigade to be performed at the workplace

The employer must ensure anyone in the fire brigade is physically able to perform their expected duties. Employers shall not permit employees to be part of a fire brigade if they have known heart disease, epilepsy, or emphysema unless a doctor gives written approval.

The employer is responsible for providing training and education to employees who are part of a fire brigade. This training must be appropriate based on employees' expected job functions under the brigade. Training and education must be provided free of cost to the employee before they are assigned duties under the brigade. Fire brigade leaders, captains, and training instructors shall be provided with training and education that are more comprehensive as compared to general members of the brigade. Training must be provided frequently enough to ensure members are able to perform their duties safely and must be provided at least annually. If brigade members are expected to perform interior structure firefighting, then they shall be provided with training at least quarterly. The quality of training for fire brigade members must be similar to those conducted by fire training schools.

In addition to firefighting training, fire brigade members shall be educated about the specific fire hazards in the workplace such as flammable liquids, toxic chemicals, water-reactive substances, and so on. If anything changes regarding these hazards, fire brigade members must be made aware. The employer must maintain written procedures for the fire brigade regarding the handling and safety precautions of these special or specific hazards.

The employer must provide, maintain, and annually inspect firefighting equipment that a fire brigade may use. Portable fire extinguishers and respirators used by the brigade must be inspected monthly in accordance with their respective, specific OSHA standards. Firefighting equipment that is damaged or defective must be removed from service until repaired or replaced. Employers must also provide fire brigade employees with effective protective clothing and equipment that protects the head, body, and extremities from fire hazards. This includes at least the following:

- Foot and leg protection
  - Achieved by either boots that protect the legs or protective footwear in combination with adequately protective pants
- Body protection (e.g., fire-resistive coat)
  - Tearing strength of outer shell a minimum of 8 pounds
- Hand protection
  - Gloves or glove systems that provide protection against cuts, punctures, and heat
- Head protection
  - Protective head device (e.g., hard hat) with ear flaps and chin strap
- Eye and face protection

- ○ To protect from the hazard of flying or falling materials (e.g., ANSI Z87 safety glasses, face shields, or full face of a respirator)
- Respiratory protection
  - ○ All respirators (e.g., self-contained breathing apparatus) provided for emergency use of a fire brigade members must comply with 29 CFR 1910.134

PORTABLE FIRE EXTINGUISHERS

Selection and Distribution

OSHA sets requirements for the placement, use, and maintenance of portable fire extinguishers. The general requirement is the employers shall provide portable fire extinguishers, mount, locate, and identify them so they are readily accessible to employees IF they allow any employee to use them. The following situations are fully or partially exempt from the OSHA Portable Fire Extinguisher standard:

- It is possible that fire extinguishers do not have to be present in a workplace. It is possible that OSHA standards that would require access to fire extinguishers (e.g. welding) do not apply to the employer. It is also possible that local building and fire codes do not require fire extinguishers in a specific building. All said, the employer may elect to enforce a policy that employees must evacuate and not attempt to extinguisher a fire in the event a fire occurs. A total absence of fire extinguishers in the workplace is uncommon but possible.
  - ○ Example: The employer is an office building. Local fire/building code does NOT require the installation of fire extinguishers if there is an automatic sprinkler system installed. There is. The employer has an Emergency Action Plan that directs employees to evacuate immediately upon noticing a fire and prohibits them to attempt to extinguish it. No portable fire extinguisher requirements apply to the employer.
- Extinguishers are available, but the employer directs employees NOT to use them and immediately evacuate upon noticing a fire. This must be documented in an Emergency Action Plan. The available extinguishers simply have to be inspected and maintained per the OSHA standard, and no other requirements apply.
  - ○ Example: The employer is a manufacturing plant with limited fire hazards. The employer has a written Emergency Action Plan that directs employees to immediately evacuate upon noticing a fire without trying to extinguish it. The extinguishers do not have to be adequately selected

and distributed, since employees are not expected to use them. However, extinguishers must be properly inspected and maintained (monthly and annually).

- When extinguishers are available but the employer only allows specific, trained employees to fight fires (e.g., fire brigade) and all other employees must evacuate immediately, then the employer is exempt from placement and distribution requirements of the fire extinguisher standard. This policy must be documented in the employer's Emergency Action Plan.
    - Example: The employer is a production facility with an area that has a specific high fire risk. The employer has an Emergency Action Plan that details the organization has a trained fire brigade. Only the fire brigade is allowed to fight fires, and all other employers are directed to evacuate immediately in the event of a fire. The extinguishers do not have to be adequately distributed; they might be stored in a designated area for the fire brigade to access and use. The extinguishers must be properly inspected and maintained.

OSHA does not explicitly require fire extinguishers to be present in the workplace, unless required by a specific standard (e.g., welding/cutting) or if an employer chooses to direct or allow employees to use fire extinguishers for fire protection. However, fire extinguishers are required by most local buildings and fire codes in most industrial settings and workplaces. It is highly recommended to assume fire extinguishers are required in your workplace and follow all provisions of the OSHA standard relative to selection, distribution, inspection, maintenance, and training.

Where the employer provides extinguishers in the workplace AND allows employees to use them, the employer shall mount, locate, and identify extinguishers so they are readily accessible (accessed with little effort, including travel distance to access). Extinguishers provided in the workplace may not be obstructed by the storage of items in front of them. Fire extinguishers are classified by their type based on the type of fire they are capable of extinguishing without presenting an additional hazard. Figure 12.1 describes different classes of fires. Portable fire extinguishers must be labeled with their class to indicate what type of fires they are appropriate for as seen in figure 12.2. The employer must ensure the proper class of extinguisher is available based on the materials and hazards in the workplace, if they allow employees to use them.

Portable fire extinguishers that are allowed to be used by any employee to fight Type A fires (ordinary combustibles) must be arranged and mounted such that the travel distance for employees to any extinguisher is 75 feet or less. Portable fire extinguishers that are allowed to be used by any employee to fight Type B fires (flammable liquids) must be arranged and mounted such

A		Ordinary Combustibles	Wood, Paper, Cloth, Etc.
B		Flammable Liquids	Grease, Oil, Paint, Solvents
C		Live Electrical Equipment	Electrical Panel, Motor, Wiring, Etc.
D		Combustible Metal	Magnesium, Aluminum, Etc.
K		Commercial Cooking Equipment	Cooking Oils, Animal Fats, Vegetable Oils

Figure 12.1

Figure 12.2

that the travel distance for employees to any extinguisher is 50 feet or less. Portable fire extinguishers that are allowed to be used by any employee to fight Type C fires (electrical equipment) must be arranged and mounted such that the travel distance for employees to any extinguisher is 75 feet or less (if the only other hazards existing are Class A fires) and 50 feet or less (if there are Class B fire hazards existing). Portable fire extinguishers that are allowed to be used by any employee to fight Type D fires (combustible metal) must be arranged and mounted such that the travel distance from any combustible metal working area to any extinguisher is 75 feet or less.

INSPECTION AND MAINTENANCE

Employers must ensure all provided fire extinguishers in the workplace are fully charged, maintained, and in operable condition. The employer must visually inspect all fire extinguishers provided in the workplace monthly to ensure they remain in operable condition and are fully charged. This inspection must be documented (e.g., on tag affixed to the extinguisher). The employer must ensure all provided fire extinguishers are subjected to an annual maintenance check. This maintenance check must be conducted in accordance with the manufacturer's instructions and is typically performed by an outside service. Annual maintenance checks must be documented (e.g., tag affixed to extinguisher). Certain, pressurized fire extinguishers must be subject to a hydrostatic test based on the test intervals found in appendix A of this chapter. A hydrostatic test evaluates the fire extinguisher vessel for integrity and pressure loss. Hydrostatic tests must be conducted by trained, qualified persons in approved testing facilities. Annual fire extinguisher maintenance is normally conducted by a fire extinguisher service organization.

FIRE EXTINGUISHER TRAINING

Where the employer has provided fire extinguishers for employee use, the employer shall provide training to employees who might use them. This shall include information about how to use fire extinguishers provided and the

Figure 12.3

hazards involved with fighting small fires. In general, employees are taught the acronym PASS. This is explained in figure 12.3. Training for those allowed or expected to use an extinguisher shall be provided annually. Those employees designated to use firefighting equipment as part of an Emergency Action Plan shall be trained on the appropriate use of the equipment they are expected to use.

Automatic Sprinkler Systems

Automatic sprinkler systems are required based on the application of specific OSHA standards and presence of specific fire hazards in the workplace. Probably more relevant and important, local fire codes may require automatic sprinkler systems to be installed in facilities based on their type of occupancy and type of industry. Where automatic sprinklers are installed, they shall be designed to provide adequate discharge patterns, water densities, and water flow characteristics for complete coverage in the workplace. There are engineering calculations for designing adequate sprinkler system coverage, but these are not covered in this book.

Employers must ensure installed automatic sprinkler systems are subject to a main drain flow test annually. This test is mainly to ensure adequate flow through the sprinkler system. The employer must also ensure the sprinkler system's inspector's test valve is opened at least every two years to ensure the system is working overall including audio and visual alarms. Figure 12.4 illustrates some key components of a typical sprinkler system.

Sprinkler systems, in general, are classified as either wet-pipe or dry-pipe systems. Wet-pipe systems contain water at all times. If any sprinkler head is opened, water will be released. Wet-pipe systems are sensitive to conditions where water can freeze in the pipes. Dry-pipe systems contain air under pressure

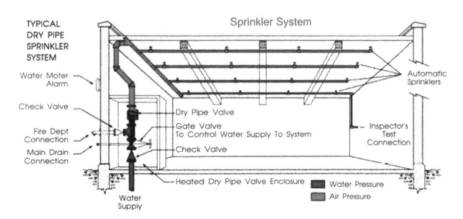

Figure 12.4

(e.g., figure 12.4). When a sprinkler head is opened, pressure is reduced which allows a valve to open and water to be released into the sprinkler system and be discharged by sprinkler heads. Most sprinkler heads are opened by heat. This can be achieved by a metal component of the sprinkler head melting from heat and allowing the sprinkler head to open, or this can be achieved by a pressure in a bulb rising from heat causing the bulb to rupture thus allowing the sprinkler head to open. If a sprinkler system has more than twenty sprinkler heads, then there must be a local water flow alarm that is sounded when water flows through the system. The employer shall ensure sprinkler systems are protected from damage. The minimum vertical clearance between a sprinkler head and the material below is 18 inches. On a related note, International Fire Code (adopted as law by 42 states) requires a minimum clearance of 24 inches between an unsprinklered ceiling and the material below.

FIRE DETECTION SYSTEMS AND ALARM

Fire detection and alarm systems are required by specific OSHA standards or when specific fire hazards are found in the workplace. More importantly, local fire codes will likely govern the presence of building-installed fire detection and alarm systems. Where installed, employers shall ensure fire detection and alarm systems remain in operable conditions. Employers must ensure fire detection and alarm system are tested and adjusted as needed to maintain proper reliability and operating conditions. Fire alarm systems may be triggered by fire detectors, may be activated by human action (e.g., pull stations), or may be an employee alarm system (e.g., employee using devices like air horns to signal an emergency). Figure 12.5 shows a typical fire alarm pull station. Access to fire alarm systems may not be obstructed. Fire detectors shall be cleaned of dirt, dust, and so on to remain operable. Fire detectors may work by detecting heat level, rate of rise of heat, or perhaps smoke. Employers must ensure corrosive or damaging chemicals do not compromise fire detector functionality. If fire detectors are used to trigger fire protection systems (sprinklers), the employer must ensure such detection triggers the system in time to adequately control or extinguish a fire.

REVIEW QUESTIONS

1. Are employers required to provide fire extinguisher to employees? If provided, do all employees have to use a fire extinguisher in the event of a fire?

Figure 12.5

2. What are the types of fires/fire extinguishers? What hazards constitute each type?
3. What is a fire brigade? What must a fire brigade policy contain? What type of training is needed for members of a brigade?
4. What are the inspection and maintenance requirements for a portable fire extinguisher?
5. What is the minimum clearance between a sprinkler head and storage below?
6. What type of tests does OSHA require to be done on a sprinkler system?
7. What are two types of sprinkler systems? What are two types of sprinkler heads?

REFERENCES

Occupational Safety & Health Administration [OSHA]. (1998). Regulations (Standards-29 CFR 1910.155). Retrieved from https://www.osha.gov/laws-regs/regulations/standardnumber/1910/1910.155

Occupational Safety & Health Administration [OSHA]. (2008). Regulations
(Standards-29 CFR 1910.156). Retrieved from https://www.osha.gov/laws-regs/
regulations/standardnumber/1910/1910.156

Occupational Safety & Health Administration [OSHA]. (2002). Regulations
(Standards-29 CFR 1910.157). Retrieved from https://www.osha.gov/laws-regs/
regulations/standardnumber/1910/1910.157

Occupational Safety & Health Administration [OSHA]. (1981). Regulations
(Standards-29 CFR 1910.159). Retrieved from https://www.osha.gov/laws-regs/
regulations/standardnumber/1910/1910.159

Occupational Safety & Health Administration [OSHA]. (1980). Regulations
(Standards-29 CFR 1910.164). Retrieved from https://www.osha.gov/laws-regs/
regulations/standardnumber/1910/1910.164

APPENDIX A—HYDROSTATIC TEST INTERVALS (OSHA TABLE L-1)

Type of extinguishers	Test interval (years)
Soda acid (soldered brass shells) (until 1/1/82)	(1)
Soda acid (stainless steel shell)	5
Cartridge operated water and/or antifreeze	5
Stored pressure water and/or antifreeze	5
Wetting agent	5
Foam (soldered brass shells) (until 1/1/82)	(1)
Foam (stainless steel shell)	5
Aqueous Film Forming foam (AFFF)	5
Loaded stream	5
Dry chemical with stainless steel	5
Carbon Dioxide	5
Dry chemical, stored pressure, with mild steel, brazed brass or aluminum shells	12
Dry chemical, cartridge or cylinder operated, with mild steel shells	12
Halon 1211	12
Halon 1301	12
Dry powder, cartridge or cylinder operated with mild steel shells	12

[1] Extinguishers having shells constructed of copper or brass joined by soft solder or rivets shall not be hydrostatically tested and shall be removed from service by January 1, 1982. (Not permitted)

Chapter 13

Powered Industrial Trucks

INTRODUCTION AND SCOPE

In industrial settings, sometimes large fork trucks, tractors, and similar vehicles are needed to move and handle heavy or hazardous material. These vehicles, typically referred to as trucks, of course have a power source to operate (e.g., combustion engine, battery). They are also used in industrial environments; therefore, we call them Powered Industrial Trucks (PITs). The most common PIT is a fork truck or forklift as seen in figure 13.1.

PITs must be used safely and maintained in a way that meets OSHA requirements and prevents incidents. PITs must meet specific American design standards. PIT operators must be trained and evaluated so they can operate safely. OSHA requires some specific behavior when it comes to operating PITs. PITs must be kept in safe working order, so these machines must be properly maintained and receive daily inspections. Finally, because some workplaces can contain flammable dust and vapors, PITs must be designated and selected based on their ability to prevent the ignition of flammable substances in certain locations. PITs are large, heavy, powerful machines, so they must be operated carefully to prevent serious incidents.

The OSHA standards covered in this chapter include:

- 29 CFR 1910.178—Powered industrial trucks

NARRATIVE

As a high school teenager, Ryan is an A-student, captain of the basketball team, sings in the chorus, and gets along with everyone. After Ryan graduates,

Figure 13.1

he attends a local college and majors in Pre-Pharmaceutical Studies. Ryan is well-liked, smart, and loved by his friends and family. During college, Ryan studies hard, has fun, and comes home every once in a while to visit his mom. During school breaks, Ryan works for a family friend who manages a warehouse. It is an unassuming warehouse building. No one expected an incident inside that building to ruin a family and stun a small town.

It is the end of the college school year, and Ryan finishes his last final exam of the spring semester. He packs his stuff, says goodbye to his roommates, and jumps in the car to drive to his parent's house. When he arrives, as usual, his mom meets him at the front door and gives him a hug before he can even set down his bags. Ryan spends the night with his mom, dad, and sister. The family has plans to go out to breakfast the next morning and celebrate Ryan completing another year of college. Ryan reluctantly informs them he has plans to work first thing in the morning at the warehouse, and he will have to take a rain check.

The next morning, Ryan jumps out of bed and drives to the warehouse he has been working at since high school. Just like Ryan, most of the employees who work there are young. The manager is a kind person and expects Ryan and the employees to simply be on time, work hard, and have fun while they are there. The warehouse is part of a small business that receives car parts, keeps inventory, stores them on warehouse shelves, and ships them to suppliers. There are a couple of full-time employees who mostly operate the forklifts. Ryan works as a stocker moving boxes, taking inventory, sweeping the floor, and helping out in anyway he can.

As far as safety at the warehouse, anyone operating a forklift needs to be "trained" by the manager. However, this "training" consists of the manager showing the operator how to work the forklift, telling them to be careful, and advising them not to goof around. The same forklifts have been used for more than a decade. The operators are not trained in identifying deficiencies of the forklift, and there is no routine maintenance to make sure they are kept in safe, working order.

That morning, Ryan is taking inventory while a forklift operator is lifting a large pallet of car batteries to the top shelf of the rack where Ryan is working. Each battery weighs about 45 pounds. The pallet is stacked 6 feet high with these heavy batteries. Some are in boxes. Some are not. The batteries are not evenly stacked. As the operator raises the load to the very top shelf, Ryan thinks he should probably get out of the way and stand clear. Ryan starts to walk under the load to head to the other end of the rack. At the same time, while the load is being raised, the chains shift from the wheels of the lifting mechanism. This causes the load to shift and several batteries fall about 30 feet from the top of the stack. As Ryan is walking under the load, one of the batteries strikes him on the top of his head. Ryan hits the concrete, unconscious, bleeding from a fracture in his skull.

A half hour later, after the incident, Ryan's mother picks up the phone to hear her son is in critical condition in the hospital. Ryan is in a coma. Over the next two weeks, Ryan's mother spends most of the day with her head on her son's chest praying he will pull through. But, on a sunny spring day, in a hospital room, Ryan's heart stops beating.

The fatal injury Ryan experienced was related to unsafe work practices at the warehouse. The company did not have a formal forklift safety program for ensuring operators were trained on inspecting the forklift for mechanical deficiencies. The forklifts received no routine maintenance to ensure deficiencies were corrected before being used. This contributed to the death of a good-hearted young man.

The line at Ryan's funeral was wrapped around the block of that small town and the service lasted well into the night. The last sight that Ryan's friends and family have to remember him by is his body in a casket, his mother's head on his chest, his shirt soaked from her tears. Ryan's family was never the same continuing life without him.

SUMMARY OF OSHA STANDARD

PIT Design

PITs must meet design requirements established by the ANSI standard for Powered Industrial Trucks (ANSI B56.1-1969). It is possible PITs built in other countries may not be built to these design and safety standards. Employers need to make sure they only use PITs that meet the ANSI standard for safe design. After being designed and manufactured, PITs cannot be altered or modified in a way that may affect their capacity or safe operation unless the manufacturer provides written approval for the alteration. All PITs have nameplates (or data plates) that tell you the weight of the truck, its capacity, and some other information. These plates, as seen in figure 13.2, need to be legible at all times.

The capacity of a PIT shall never be exceeded. The capacity of a PIT, namely a forklift, is typically expressed as maximum weight, in pounds, at a distance from the mast. On a forklift, the mast is the vertical component or fixture that uses mechanisms to raise and lower a load. Figure 13.3 illustrates typical forklift components and shows the mast. Commonly, forklift capacity is given as a weight limit (pounds) at 24 inch from the mast. This is because typical wooden pallets are 48 inch wide. So, if the pallet load is pushed against the mast, the center of the load is 24 inch from the mast, and the capacity rating would apply. If the center of the load is more than 24 inch from the mast (or distance expressed in capacity rating), then the capacity of the machine would be reduced. Some data plates will express capacities at center-of-load locations other than 24 inch. Some manufacturers provide

Figure 13.2

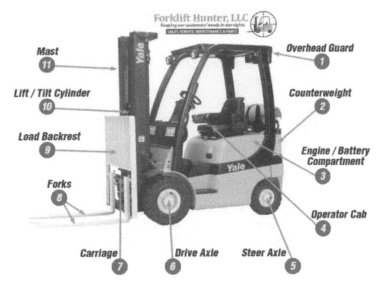

Figure 13.3

guides and charts in their manual for identifying capacities at different center-of-load locations.

OPERATOR TRAINING

The employer must make sure each PIT operator is competent to operate each PIT he or she may operate. The operator must be knowledgeable of the hazards related to the operation, the PIT itself, and how to avoid incidents. Operators must first receive training on PIT operation and must demonstrate the ability to operate safely before they can operate a PIT without direct supervision. Operators-in-training may operate a PIT with the direct supervision of someone who is an experienced, trained operator. Operators must receive a combination of instruction (e.g., classroom) training and a hands-on, guided demonstration of skill. The instructional training shall include the following PIT related topics:

- Operating instructions and precautions for types of PITs to be operated
- Differences between PITs and automobiles
- PIT controls and instrumentation (where they are and what they do)
- Engine or motor operation
- Safe steering/maneuvering
- Important safety tips on visibility

- Fork and attachment operation and limitations
- Vehicle capacity
- Stability
- Required inspection and maintenance an operator must perform
- Refueling/charging procedures
- Operating limitations
- Other operating instructions, warning, and precautions found in the PIT operator's manual
- Workplace-specific topics such as but are not limited to:

> surface conditions, information on loads to be carried, proper stacking and load manipulation, pedestrian traffic, narrow aisle or restricted operating zones, hazardous locations (e.g., locations with flammable vapors), ramps and slopes in the workplace, and other potentially hazardous environmental conditions

After receiving formal, instructional training, any operator must be evaluated on the ability to operate safely. This must consist of a trained, experienced PIT operator witnessing the operator demonstrate the ability to operate safely. Operators need to have this evaluation before they are certified to operate without direct supervision. Operators must be reevaluated every three years. A refresher instructional training is required for operators if the operator is observed operating in an unsafe manner, if their operator was involved in an incident, the operator is assigned to drive a new or different PIT, or the workplace changes that warrant the need for a refresher training.

After training and evaluation, the employer must certify PIT operators by documenting the training and evaluation process. This includes documenting the name of the operator, the date of the training, the date of evaluation, and the name of the person who performed the training and evaluation. Often, companies provide PITs licenses to operators to fulfill this certification requirement.

SAFE OPERATIONS AND TRAVEL

The following are some OSHA-required DOs and DONTs of PIT operations (see table 13.1).

Table 13.1 OSHA DOs and DONTs of PIT OPERATION

DO	Do _NOT_
• When leaving a PIT unattended, (+25 feet away), turn it off • If leaving a PIT unattended, but still within 25 feet and in view, fully lower load and set parking brake • Keep safe distance between edges of ramps or platforms • Ensure an overhead guard is in place above the operator compartment to protect against falling objects • Travel at a safe speed and keep at least three truck lengths between multiple PITs traveling the same direction in a line • Slow down and sound the horn at all cross-aisles and when visibility is obstructed • Travel in the direction of best visibility (so the load doesn't block the view) • Travel slowly on ramps and slopes • When ascending/descending a grade in excess of 10 percent, loaded PITs shall be driven with load upgrade (forks pointing uphill) • Negotiate turns slowly • Remove PITs from service if found unsafe and do not return to service until the issue is corrected	• Drive a PIT toward a person who is standing in front of a fixed object or wall • Walk under the elevated load of a PIT • Allow passengers on PITs if there is not a designed passenger seat/compartment • Place limbs in danger zones of PITs where they could be moving parts • Travel forward if you can't see what is in front of you • Horseplay, race, etc. • Run over lose objects • Handle unstable loads • Operate a truck that needs to be repaired, is defective, or is unsafe in any way • Operate a truck with any spillage or leak of oil, fuel, or fluids

MAINTENANCE AND INSPECTION

Any power-operated industrial truck not in safe operating condition shall be removed from service. All repairs shall be made by authorized personnel who know how to fix the problem properly. Any parts replaced on a PIT shall be equivalent to the original parts used in the original design.

PITs shall be inspected at least daily, on days they are used, before they are operated, to identify any unsafe conditions. If PITs are used round-the-clock, then they must be inspected before each shift. These pre-use inspections must be documented and are typically kept with the machine. If any unsafe conditions are found, they shall be removed from service immediately until the problem is corrected. Inspections should be conducted in accordance with manufacturer recommendations. Appendix A provides example of a forklift pre-use inspection.

HAZARDOUS LOCATION DESIGNATIONS

PITs must be designated by their fuel source (e.g., diesel fuel, gasoline, liquid propane, and battery/electric powered). The following list provides designations and their description:

Diesel Fuel

* D: diesel-powered PITs
* DS: diesel-powered PITs with additional safeguards to exhaust, fuel system, and electrical system (for prevention of flammable materials)
* DY: diesel-powered PITs with all safeguards as a DS designation and in addition do not have any electrical equipment including the ignition and are equipped with temperature limitation features (for prevention of flammable materials)

So, as far as using diesel-powered forklifts in locations where there might be flammable material in the air, D would be the least safe, diesel-powered option, DY would be the most safe, diesel-powered option, and DS would be between D and DY.

Gasoline Fuel

* G: gasoline-powered PITs
* GS: gasoline-powered PITs with additional safeguards to exhaust, fuel system, and electrical system (for prevention of flammable materials)
* GY: gasoline-powered PITs with all safeguards as a DS designation and in addition do not have any electrical equipment including the ignition and are equipped with temperature limitation features (for prevention of flammable materials)

So, as far as using diesel-powered forklifts in locations where there might be flammable material in the air, G would be the least safe, gasoline-powered option, GY would be the most safe, gasoline-powered option, and GS would be between G and GY.

Electric Power Source

* E: electrically powered PITs that have minimum acceptable safeguards against inherent fire hazards

- ES: electrically powered PITs that have the safeguards E has and have additional safeguards to the electrical system to prevent sparks and to limit surface temperatures
- EE: electrically powered PITs that have the safeguards ES has and the electric motor is completely enclosed
- EX: electrically powered PITs that have the safeguards EE has and are practically full-proof against igniting flammable vapors or dust

So, as far as using electric powered forklifts in locations where there might be flammable material in the air, E would be the least safe, electrically powered option and EX would be the most safe.

Liquid Propane Fuel

- LP: liquid propane-powered PITs
- LPS: liquid propane-powered PITs with additional safeguards to exhaust, fuel system, and electrical system (for prevention of flammable materials)

So, as far as using liquid propane-powered forklifts in locations where there might be flammable material in the air, LP would be the least safe, propane-powered option and LPS would be the most safe.

Why do these designations matter? Well, some industrial environments can contain flammable vapors, dusts, or fibers in the air. The employer must assess the hazards and risk of igniting airborne flammables when selecting types (designations) of PITs that will be used in the workplace. OSHA gives some specific guidance and prohibits PITs from being used in certain atmospheres. For example, PITs shall not be operated in atmospheres containing hazardous concentrations of acetylene and hydrogen, which are very flammable materials. Only PITs designated EX may be used in atmospheres containing hazardous concentrations of acetone, metal dust, alcohol, natural gas, and some other substances. It is up to the employer to assess the hazard and determine if there is potential for hazardous concentrations of materials in the air that may be ignited. OSHA advises only PITs designed as DY, EE, or EX may be used in locations where flammable liquids are processed, but the vapors or gases are confined in closed containers during normal operations. OSHA provides more specific examples and prohibitions in the PIT standards; however, it is ultimately up to the employer to select the right designation of forklift based on hazard assessment of the environments they will be operated in.

Figure 13.4

PIT BATTERY CHARGING AREAS

Any areas reserved for the charging of PIT batteries must be designated for that purpose only. Figure 13.4 gives an example of signage. There must be supplies in the designated area to clean up any spilled or leak electrolyte/acid from batteries. There must also be measures for fire protection (e.g., sprinklers, fire extinguisher) in the area. Measures must be taken to prevent charging cords and devices from being run over or damaged by trucks. The area must have adequate ventilation (e.g., some batteries can release fumes like hydrogen). PIT batteries are likely very heavy, so hoists or similar devices shall be provided if heavy batteries must be handled. When charging, it shall be required that PITs are turned off with the brake applied. Smoking shall be prohibited in designated charging areas and measures must be taken to prevent open flames, sparks, or other ignition sources.

If employees must service (i.e., add water) to forklift batteries that contain acid, then certain precautions are required. This work would require appropriate PPE including a faceshield, acid resistant gloves, and an acid resistant apron. Also, this work would require access to an emergency eyewash and safety shower that is in the work area available for immediate use.

REVIEW QUESTIONS

1. Describe the training and evaluation process for PIT operators.
2. What is a PIT data plate a.k.a. nameplate?
3. Name six topics that must be part of the instructional portion of a PIT training program.
4. What must an operator do if leaving a PIT unattended (+25 feet away)?
5. An operator is diving a loaded forklift down a hill that is an 11 percent grade. The load is a full-size pallet stacked 6 feet high with boxes. In what direction must the driver descend down the hill? Give two reasons for your answer.
6. How often are PITs required to receive a safety inspection?
7. What designations of forklifts are allowed to be used in a location that contains processes that use and store flammable liquids in which flammable gases and vapors are contained in enclosed systems? What designation of forklift would be the best choice for operation in an environment that regularly contains hazardous concentrations of metal dust?

REFERENCES

Occupational Safety & Health Administration [OSHA]. (2016). Regulations (Standards-29 CFR 1910.178). Retrieved from https://www.osha.gov/laws-regs/regulations/standardnumber/1910/1910.178

"Sample Daily Checklists for Powered Industrial Trucks." *Training and Reference Materials Library*, Occupational Safety & Health Administration [OSHA]. www.osha.gov/training/library/powered-industrial-trucks/checklist.

APPENDIX A—FORKLIFT PRE-USE INSPECTION CHECKLIST (OSHA TRAINING LIBRARY)

Operator's Daily Checklist - Internal Combustion Engine Industrial Truck - Gas/LPG/Diesel Truck

Record of Fuel Added

Date	Operator	Fuel
Truck#	Model#	Engine Oil
Department	Serial#	Radiator Coolant
Shift	Hour Meter	Hydraulic Oil

SAFETY AND OPERATIONAL CHECKS (PRIOR TO EACH SHIFT)
Have a **qualified** mechanic correct all problems.

Engine Off Checks	OK	Maintenance
Leaks – Fuel, Hydraulic Oil, Engine Oil or Radiator Coolant		
Tires – Condition and Pressure		
Forks, Top Clip Retaining Pin and Heel – Check Condition		
Load Backrest – Securely Attached		
Hydraulic Hoses, Mast Chains, Cables and Stops – Check Visually		
Overhead Guard – Attached		
Finger Guards – Attached		
Propane Tank (LP Gas Truck) – Rust Corrosion, Damage		
Safety Warnings – Attached (Refer to Parts Manual for Location)		
Battery – Check Water/Electrolyte Level and Charge		
All Engine Belts – Check Visually		
Hydraulic Fluid Level – Check Level		
Engine Oil Level – Dipstick		
Transmission Fluid Level – Dipstick		
Engine Air Cleaner – Squeeze Rubber Dirt Trap or Check the Restriction Alarm (if equipped)		
Fuel Sedimentor (Diesel)		
Radiator Coolant – Check Level		
Operator's Manual – In Container		
Nameplate – Attached and Information Matches Model, Serial Number and Attachments		
Seat Belt – Functioning Smoothly		
Hood Latch – Adjusted and Securely Fastened		
Brake Fluid – Check Level		

Engine On Checks – Unusual Noises Must Be Investigated Immediately	OK	Maintenance
Accelerator or Direction Control Pedal – Functioning Smoothly		
Service Brake – Functioning Smoothly		
Parking Brake – Functioning Smoothly		
Steering Operation – Functioning Smoothly		
Drive Control – Forward/Reverse – Functioning Smoothly		
Tilt Control – Forward and Back – Functioning Smoothly		
Hoist and Lowering Control – Functioning Smoothly		
Attachment Control – Operation		
Horn and Lights – Functioning		
Cab (if equipped) – Heater, Defroster, Wipers – Functioning		
Gauges: Ammeter, Engine Oil Pressure, Hour Meter, Fuel Level, Temperature, Instrument Monitors – Functioning		

Chapter 14

Cranes and Slings

INTRODUCTION AND SCOPE

Cranes are useful machines in industrial environments where heavy or otherwise hazardous materials must be handled and transported. Cranes provide an alternative to lifting and moving materials that would be hazardous for employees to handle manually. However, cranes and associated equipment must be used safely, so operation does not create a hazard. Unsafe use of cranes can result in large, heavy material being dropped on people or equipment. While cranes can come in different shapes and sizes, this chapter focuses on overhead bridge cranes and gantry cranes.

Overhead and gantry cranes are the most common types of cranes in industry. Overhead cranes have an overhead bridge that sits on overhead rails. The bridge moves the crane (e.g., north and south) along its runway. A hoisting mechanism is attached to the bridge. A hoist moves the load up and down (lifting and lowering). A hoist can usually be moved back and forth (e.g., left to right, or east to west) on the bridge by means of a trolley. Gantry cranes are similar to overhead cranes, only instead of overhead rails supporting the bridge, the bridge is supported by upright legs (gantries). Most commonly, cranes are capable of providing powered movement to a load in three directions (north and south by moving the bridge, east and west by moving the trolley or hoist on the bridge, and up and down by lifting and lowering with the hoist). The OSHA standard summarized in this chapter applies to only overhead and gantry cranes. It does not apply (in full) to simple hoists that are fixed on a beam or that even may be underhung, rolled, and be moved with manual force on a beam or rail system. There are a few paragraphs that address hoisting equipment.

Cranes must be constructed based on design standards for safety. Cranes must have adequate clearance surrounding their runway, and precautions must be taken so personnel on the ground are not endangered by movement of the crane. Only trained and qualified persons may operate a crane. Cranes shall be equipped with certain safety features that prevent the crane from running into things or crane components from running into each other. Hoists of cranes must be equipped with adequate braking mechanisms to support a load based on its rated capacity. The capacity of a crane must be plainly posted, so loads weighing more than its rated capacity are not picked up. Loads shall be carried in a way to avoid lifting above people, avoid swinging of the load, and otherwise to maintain safety.

Beyond safe operation, cranes must be properly inspected and maintained to keep them free from hazardous conditions. Cranes should receive a visual inspection before each use to ensure there are no conditions that may compromise safety. Certain crane components like hooks and chains shall receive a documented, monthly inspection. A crane must receive a comprehensive, documented inspection at least annually. After a crane is installed, before initial use, it must be load tested to ensure it can withstand its rated capacity and then some.

Cranes use slings to pick up loads. Slings are arrangements of materials and equipment (rigging) that attach the load to the crane's hook. Slings must be arranged to ensure the rigging will not be overloaded and the load will be picked up in a balanced, safe manner. Common rigging materials include metal chains, wire ropes, and synthetic (e.g., nylon) straps. Rigging must be inspected before each use to ensure it is safe to use. OSHA outlines specific criteria that, if observed, require rigging to be removed from service. All rigging must have a tag that indicates its capacity based on the sling configuration.

The OSHA standards covered in this chapter include:

- 29 CFR 1910.179—Overhead and gantry cranes
- 29 CFR 1901.184—Slings

Not included:

- 29 CFR 1910.180—Crawler locomotive and truck cranes
- 29 CFR 1910.181—Derricks
- 29 CFR 1910.183—Helicopters

NARRATIVE

Sawyer is a crane operator at a steel mill. His supervisor is known to be hard to work for. The supervisor is impatient and pushy. While the company has a pretty extensive safety program, Sawyer's supervisor is known to disregard safety procedures. His supervisor has been working in the steel mill since before OSHA was created. He tends to think safety gets in the way of things. Sawyer is the complete opposite. He respects the rules and thinks safety is very important. He would never want to see anyone get hurt.

The steel mill is going through some major changes and rearranging. Each day, Sawyer typically operates an overhead crane from a cab to pick up and move large rolls of steel. Because of the changes to the mill, instead of moving rolls of steel, Sawyer's supervisor tells him he will be asked to move a large piece of equipment from one location to another. Sawyer is very familiar with the rolls of steel he typically lifts with the crane. He knows how much they weigh and how to pick them up safely. However, the equipment Sawyer is asked to move with the crane is much larger than the normal loads of steel. There is no telling how much it weighs. Also, it is unclear how the equipment will be rigged and safely picked up by the crane. Sawyer expresses concern to his supervisor. Sawyer says he is not comfortable using the crane to move the equipment without knowing at first how much it weighs. The last thing Sawyer wants is to be blamed for dropping the equipment or for someone getting hurt.

It is time to move the equipment, and Sawyer's supervisor demands that he get in the cab and move the equipment to its new location. It takes all of Sawyer's courage, but he tells his supervisor he will not do it unless someone can at least prove to him how much it weighs. Sawyer's supervisor yells at him to go home and that he will find someone else to do it who has the guts to get things done. Sawyer does not argue and leaves the mill with pride of his decision.

The replacement crane operator is not as skilled as Sawyer when it comes to crane operation, but he climbs in the cab to move the equipment to impress the supervisor. The operator starts to pick up the equipment. The large piece of equipment is connected to the crane awkwardly. It takes five men on the ground to push and manipulate the load to keep it straight as the operator moves it. Because the operator is not very comfortable, he stops the crane rather abruptly. This causes the load to shift. The shifting load put stress on the already overloaded cables used to pick up the load. The load is much too heavy for the rated capacity of the cables used to pick it up. With the shifting of the load, one of the cable snaps, and the load comes crashing to the floor.

As the load falls to the floor, it does so with impressive force and momentum. One of the employees who was on the ground attempting to guide the equipment fell as the load shifted. As a result, the enormously heavy piece of equipment fell on to the legs of the fallen employee. The employee's screams echo throughout the mill. The employee is pinned by his now-crushed legs under the load, and he remains trapped until new cables can be retrieved and the load can be picked up again by the crane. Once new ropes are attached to the load, it finally can be lifted from the employee. There is no question the employee will be able to use his legs again. In fact, his legs were amputated at the hospital, because there was no saving them. Sawyer's friend texted him about the incident. Sawyer was proud he made a decision to not be part of the situation, but the feelings of grief and sympathy for his coworker are overwhelming.

SUMMARY OF OSHA STANDARDS

Definitions

It may be helpful to refer to these definitions as you proceed through this chapter. Major components of an overhead crane are pointed out in figure 14.1.

- Crane—a machine for lifting and lowering a load and moving it horizontally, with the hoisting mechanism being an integral part of the machine
- Block (load block)—the assembly of hook or shackle, swivel, bearing, sheaves, pins, and frame suspended by the hoisting rope
- Bridge—part of a crane consisting of girders, trucks, end ties, foot walks, and a drive mechanism that carries the trolley or trolleys

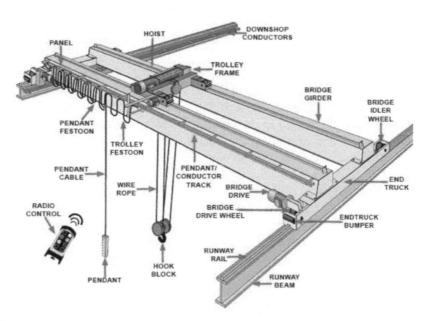

Figure 14.1

- Drum—the cylindrical member around which the ropes are wound for raising or lowering the load
- Hoist—an apparatus that may be a part of a crane, exerting a force for lifting or lowering
- Runway—an assembly of rails, beams, girders, brackets, and framework on which the crane or trolley travels
- Slings and rigging—materials and equipment configured or arranged between the load and hook of a crane; chains, wire rope, and synthetic straps used to pick up a load
- Trolley—a wheeled mechanism from which a hoist is suspected; provides horizontal motion of the hoist (e.g., along the bridge)

General Requirements

A crane is a machine for lifting and lowering a load and moving it horizontally, with the hoisting mechanism (the parts that do the lifting and lowering) being an integral part of the machine. Two of the most common cranes in industrial settings are overhead cranes and gantry cranes. Overhead cranes have a movable bridge that carries a movable or fixed hoist. The bridge travels along an overhead fixed runway structure. Figure 14.2 depicts an overhead crane. A gantry crane is similar to an overhead crane except the bridge is

Figure 14.2

supported on two or more legs that run on fixed rails or a runway. Figure 14.3 depicts a gantry crane.

All overhead and gantry cranes constructed and installed shall meet the specification of the ANSI B30.20-1967. This ensures cranes used in industry are designed safely. If a crane is modified in any way that may impact a crane's capacity, the crane shall undergo a new load rating test in accordance with a qualified engineer or the equipment manufacturer. This ensures the crane's lifting capacity is not changed or compromised. All cranes must be marked to visually identify their capacity (load rating). The load rating must be plainly marked on each side of the crane and on each hoisting unit (if more than one). It must be able to be read from ground level. A minimum of 3 inches overhead (above) and 2 inches laterally (beside) shall be provided and maintained between a crane and any obstructions (other parts/objects of the facility). Walkways for personnel shall be provided such that the safety of personnel on the ground will not be jeopardized by movements of the crane. If the runway of two cranes are parallel (next to each other), and there are no walls or structures between them, there must be adequate clearance between the two bridges of the cranes to make sure they will not run in to each other.

Only designated and qualified personnel are permitted to operate a crane. OSHA does not set specific training requirements for overhead crane and

Figure 14.3

gantry crane operators. It is up to the employer to ensure operators are qualified. OSHA defines a qualified person as:

> One who, by possession of a recognized degree, certificate, or professional standing, or who by extensive knowledge, training, and experience, has successfully demonstrated his ability to solve or resolve problems relating to the subject matter, the work, or the project.

Overhead Crane Cabs

For the most part, cranes are either floor-operated or cab-operated. If floor-operated, the operator uses a pendant (controller attached by an electrical conductor to the crane) or radio controller to move the crane. Figure 14.4 shows an operator using a pendant-controlled overhead crane. Some cranes have an operator cab where the operator sits to control the crane like the one seen in figure 14.5. These cabs are located on the bridge or trolley. All operating controls in the cab must be within convenient reach of the operator when facing the area to be served by the crane's load hook or while facing the direction of travel of the cab. The operator must have a full view of the load hook in all cab positions. The cab shall be positioned to afford a minimum

Figure 14.4

Figure 14.5

of 3 inches of clearance from all fixed structures next to the crane. Access to the cab of a crane must be by a conveniently placed ladder or stairway that does not require the operator to step over a gap exceeding 12 inches. Elevated walkways operators may use to access a crane's cab must provide at least 48 inches of headroom and must be designed to sustain at least 50 pounds per square inch. Walkways must be an anti-slip surface. All OSHA requirements concerning elevated walkways, fall protection, stairways, and fixed ladders apply to potential access methods to crane cabs.

Crane Safety Features

Overhead cranes are required to have certain safety features. Overhead cranes shall be provided with trolley stops that limit the travel of the trolley so as to not create a hazard (prevent a trolley from being run off a bridge or crash). Trolley stops shall be designed and fastened so as to resist the force that may be applied to them when contacted. In general, most overhead cranes must be provided with bridge bumpers or other effective means capable of stopping the crane if it will operate near the ends of the runway. The bumpers shall be capable of stopping the crane (not including the lifted load) at an average rate of deceleration not to exceed 3 feet/s/s when traveling in either direction at 20 percent of the rated load speed. In general, most overhead cranes must be provided with trolley bumpers or other effective means capable of stopping the trolley if it will be operated near the end of the trolley's travel (bridge). The bumpers shall be capable of stopping the trolley (not including the lifted load) at an average rate of deceleration not to exceed 4.7 feet/s/s when traveling in either direction one-third of the rated load speed. When more than one trolley is operated on the same crane's bridge, bumpers shall be installed that prevent the trolleys from running into each other. Bridge or trolley bumpers are some energy-absorbing devices to reduce or prevent impact when a bridge or trolley comes to the end of its permitted travel. (They prevent things from running in to each other or crashing into the walls of the facility.) In general, as long as cranes are designed and installed by qualified professionals, and meet the ANSI design standard, safety and health professionals do not typically evaluate these safety features.

Any exposed moving parts of a crane such as gears, chains, chain sprockets, and reciprocating components that might constitute a hazard shall be guarded with securely fastened guards.

Each hoist on a crane shall be equipped with a holding brake applied directly to the motor shaft or part of the hosting mechanisms gear train. A holding brake automatically prevents motion when power is off to the crane or hoist. In general, all crane hoists shall also be equipped with control braking means. This is a braking means by which the motor speed is controlled when an overloaded condition occurs.

There are some requirements when it comes to brakes for trolleys and bridges. Foot-operated brakes for these components shall not require an applied force of more than 70 pounds to develop the manufacturer's rated brake torque. Brake pedals shall be constructed so the operator's foot will not easily slip off the pedal. Foot-operated brake pedals shall be equipped with automatic means for positive release to come back to its original position when pressure is released from the pedal. Brakes shall be an adequate size to stop the trolley or bridge within a distance in feet equal to 10 percent of full load speed in feet per minute when traveling at full speed with full load. Again, while these safety features are required by OSHA, as long as the crane is designed and installed properly, safety professionals do not typically evaluate them.

Electrical Equipment

Electric circuits that control a crane shall not exceed 600 volts. The voltage at control pendants shall not exceed 150 volts AC or 300 volts DC. Pendants shall be constructed to prevent electrical shock and shall be clearly marked for identification of functions. Any electrical equipment shall be so located or enclosed such that live parts will not be exposed to accidental contact under normal operating conditions. In most cases, the runway rails of overhead cranes and gantry cranes are energized. This is what allows for the movement of the bridge.

Hoists and Hoisting Equipment

A hoist is a mechanism that lifts and lowers a load. Hoists may be used individually (e.g., mounted to an overhead beam) or they may be an integral part of an overhead or gantry crane. Figure 14.6 depicts a common hoist. Most hoists use wire rope. A motor of a hoist turns a drum to take up or let out the rope. Sheaves are part of the load block. Sheaves are grooved pulleys that allow for the block to be raised and lowered. Sheave grooves shall be smooth and free from defects that could cause rope damage. There must be an adequate amount of wire rope on the drum. No less than two wraps of rope shall remain on the drum when the hook is in its extreme low position. Hooks of hoists shall meet the manufacturer's recommendations and not be overloaded.

Crane Inspections

Cranes must receive initial, frequent, and periodic inspections. An initial inspection must be conducted on a crane after installation prior to initial use. The crane shall be initially inspected to ensure compliance with design standards. Frequent inspections must occur at daily to monthly intervals (typically before each use) and must be conducted by a competent person (e.g., qualified operator). Frequent inspections do not need to be documented; however, certain components require monthly, documented inspection—which is spelled out in this chapter. Periodic inspections must occur at a frequency of one month

Figure 14.6

to twelve-month intervals (normally annually). Periodic inspections must be done by a qualified person (most commonly an outside crane inspection service or technicians from the manufacturer). As a reminder, these inspections apply explicitly to overhead and gantry cranes, not simple overhead mounted hoists.

Frequent inspections (e.g., before each use) include an examination of the following:

- All functional operating mechanisms for maladjustment (each day used)
- Deterioration or leakage in lines, tanks, valves, drain pumps, and other parts of air or hydraulic systems (each day used)
- Hooks
  - Inspection for deformation or cracks (each day used)
  - Monthly documented inspection (certified with employer signature and date)
    - Hooks with more than 15 percent of normal throat opening or more than 10 degrees of twist shall be removed from service
- Hoist chains and connections
  - Visual inspection for safe conditions (each day used)
  - Month documented inspection (certified with employer signature and date)
    - Chains with excessive wear, twist, distorted links, and stretch beyond manufacturer's recommendations shall be removed from serviced (hoist not used)
- Hoist wire ropes
  - Visual inspection for safe conditions (each day used)
  - Monthly documented inspection (certified with employer signature and date)
- All functional operating mechanisms for excessive wear (each day used)

Periodic inspections (e.g., annually) must be documented (certified with employer signature and date). Periodic inspections must be conducted in accordance with manufacturer recommendations and include examination of (at minimum, for example):

- All frequent inspection examination items
- Deformed, cracked, or corroded members
- Loose bolts or rivets
- Cracked or worn sheaves and drums
- Worn, cracked, or distorted parts
- Excessive wear on brake system parts
- Load, wind, and other indicators over their full range, for any significant inaccuracies

- Excessive wear of chain drive sprockets and excessive chain stretch
- Electrical apparatus, for signs of pitting or any deterioration

If a crane sits idle (not used) for one month or more, but less than six months, it must receive a frequent inspection before use. If sitting idle for more than six months, it must receive a periodic inspection.

   Prior to the initial use of all new or modified cranes, the crane shall be tested to ensure proper function. Before initial use, a crane shall undergo a rated load test. Test loads shall not be more than 125 percent of the rated load. The load test shall be documented and available upon request. Initial use function testing includes tests of:

- Hoisting and lowering
- Trolley travel
- Bridge travel
- Limit switches and locking safety devices

A summary of crane inspection requirements is simplified and outlined below:

- Before each use (visual, not documented)
  - Minimum items outlined above and items per manufacturer's recommendations
  - Usually performed by a qualified operator
- Monthly (documented)
  - Crane wire ropes and chains
  - Hooks
  - Usually performed by a qualified operator
- Annual (documented)
  - Minimum items outlined above and items per manufacturer's recommendations
  - Usually performed by qualified service technicians

Wire Rope Inspection

All wire ropes of a crane (hoist of a crane) shall be inspected at least once a month. This inspection shall be documented (certified with employer signature and date). Any deterioration resulting in potential loss of strength or a safety hazard shall result in removing the rope or entire crane from service. Some considerations that could result in loss of strength are:

- Reduction of rope diameter below nominal diameter due to loss of core support, internal or external corrosion, or wear of outside wires.

- A number of broken outside wires and the degree of distribution or concentration of such broken wires.
- Worn outside wires.
- Corroded or broken wires at end connections.
- Corroded, cracked, bent, worn, or improperly applied end connections.
- Severe kinking, crushing, cutting, or unstranding.

Crane Maintenance

A preventative maintenance program for cranes within an employer's establishment shall be implemented based on the crane manufacturer's recommendations (typically captured by periodic inspections). Before maintenance or repairs on a crane, the following precautions shall be taken: the crane shall be positioned in an area where it will not interfere with other cranes, controllers shall be in the off position, the emergency or electrical disconnect shall be in the open position, and a warning or out of order signs shall be placed on the crane or near the hook. OSHA requirements for the control of hazardous energy (lockout/tagout) apply for service and maintenance tasks on a crane. If any unsafe condition is identified during inspection or maintenance, the crane shall be removed from service such that operation is prevented.

Load Handling

A crane shall not be loaded beyond its load capacity except for test purposes (e.g., initial load rating test). The load shall be attached to the load block hook by means of a sling or other approved devices. The load must be well secured and balanced during lifting more than a few inches. Before hoisting a load, the following conditions shall be noted:

- Multiple part lines shall not be twisted around each other.
- The hook shall be brought over the load in such a manner as to prevent swinging.
- During hoisting, care shall be taken that:
  ○ There is no sudden acceleration or deceleration of the moving load.
  ○ The load does not contact any obstructions.

Cranes shall not be used for side pulls (to pull load from the side, horizontally instead of lifting or lowering) unless a responsible, competent person has determined the stability of the crane will not be compromised. Employees may not ride on the crane hook or block. The employer shall not allow loads to be carried over people. If a load is approaching or close to the capacity of a crane, the operator shall test the brake by raising the load a few inches and applying the brakes. If two or more cranes are required to lift a load, a qualified person shall be in charge of the operation. He or she shall analyze the sling and rigging used to ensure the cranes and equipment involved are adequate for the lift. At the beginning of each operator's shift a crane is used, the upper limit switch of each hoist shall be tried-out with no load. This means the block/hook shall be inched up to the limit switch (upper limit of motion) and run at low speed. The limit switch shall cut off power before the block/hook contact the hoist/trolley. If the limit switch does not operate properly, the crane shall not be used to lift a load.

Slings

OSHA sets requirements for materials and equipment (rigging) used to handle loads lifted with a crane. The arrangement of rigging to pick up (rig) a load is called a sling. These slings are commonly composed of configurations of steel chains, wire rope, or synthetic ropes/straps. Examples of rigging are seen in figure 14.7. Common sling configurations are seen in figure 14.8. Slings can be made up of one or more legs. For example, referring to figure 14.8 vertical hitch has one leg, and a bridle hitch has two legs both at an angle. It is important to note that using slings in an angled configuration reduces the capacity of the rigging. Calculating rigging and sling capacity based on the sling configuration uses basic trigonometry. These calculations are not covered in this book.

Slings may also use under-the-hook devices. These are devices that help with arranging slings and distributing the load. A sling that uses an under-the-hook device with a gantry crane is seen in figure 14.9.

Figure 14.7

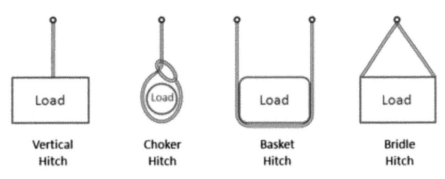

| Vertical | Choker | Basket | Bridle |
| Hitch | Hitch | Hitch | Hitch |

Figure 14.8

Figure 14.9

Safe Rigging Practices

The following is a list of safe work practices required by OSHA when using rigging and slings.

- Slings that are damaged or defective shall not be used.
- Slings shall not be shortened with knots or bolts or other makeshift devices.
- Sling legs shall not be kinked.
- Slings shall not be loaded in excess of their rated capacities.
- Slings used in a basket hitch shall have the loads balanced to prevent slippage.
- Slings shall be securely attached to their loads.

- Slings shall be padded or protected from the sharp edges of their loads.
- Suspended loads shall be kept clear of all obstructions.
- All employees shall be kept clear of loads about to be lifted and of suspended loads.
- Hands or fingers shall not be placed between the sling and its load while the sling is being tightened around the load.
- Shock loading (sudden tension on sling) is prohibited.
- A sling shall not be pulled from under a load when the load is resting on the sling.
- Employers must not load a sling in excess of its recommended safe working load as prescribed by the sling manufacturer on the identification markings permanently affixed to the sling.
- Employers must not use slings without affixed and legible identification markings.

Slings and rigging must be visually inspected before each use for damage or defects. The person performing the inspection must be competent to identify potential unsafe conditions. Damaged or defective slings shall be immediately removed from service.

Alloy Chain Slings

Alloy chain slings shall receive a documented inspection at least annually. This record of inspection must be available upon request. It is a best practice to maintain documented inspection of any and all rigging equipment including wire ropes, chains, synthetic straps, and any under-the-hook devices. Alloy steel chain slings must have a permanently affixed identification that states its size, grade, rated capacity, and length. All alloy steel chain slings must be proof tested by the manufacturer before being put into use. Worn or damaged steel chain slings shall not be used. Chain width becomes small if it becomes stressed. This is because tension over time stretches the links of the chain causing the chain links to increase in length and decrease in width. For this reason, OSHA prohibits chains from being used if their size (width) drops below a threshold. This threshold is based on its original size. If the chain size at any point of a link is less than that stated in OSHA Table N-148-1 (appendix A of this chapter), the employer must remove the chain from service. Further, the chain must be removed from service if there are any signs of cracked or deformed links.

Wire Rope Slings

Like chains used for rigging, wire rope slings must have a permanently affixed legible identification that indicates its capacity at different angles used in different types of hitches. Wire ropes slings shall be immediately removed from service if any of the following conditions are met:

- Ten randomly distributed broken wires in one rope lay, or five broken wires in one strand in one rope lay. See figure 14.10 for reference, specifically notice a lay is the length it takes for one stand to make a complete wrap around the core.
- Wear or scraping of one-third the original diameter of outside individual wires.
- Kinking, crushing, bird caging, or any other damage resulting in distortion of the wire rope structure.
- Evidence of heat damage.
- End attachments that are cracked, deformed, or worn.
- Hooks that have been opened more than 15 percent of the normal throat opening measured at the narrowest point or twisted more than 10 degrees from the plane of the unbent hook.
- Corrosion of the rope or end attachments.

Figure 14.11 provides examples of wire rope criteria that would warrant removal from service.

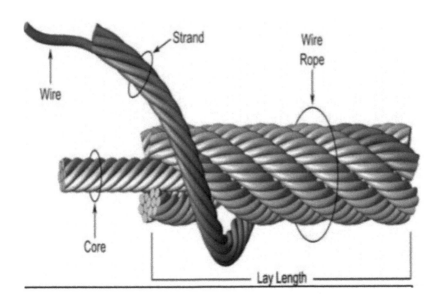

Figure 14.10

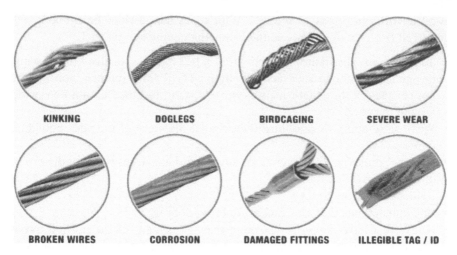

KINKING   DOGLEGS   BIRDCAGING   SEVERE WEAR

BROKEN WIRES   CORROSION   DAMAGED FITTINGS   ILLEGIBLE TAG / ID

Figure 14.11

Natural and Synthetic Fiber Rope and Web Slings

All natural and synthetic fiber rope or web slings (nylon straps being most common) must have a permanently affixed legible identification marking its rated capacity for different types of hitches that it may be used. Use of these types of slings should be avoided where there is the presence of corrosive materials and vapors. Slings made of synthetic materials shall be immediately removed from service if any of the following conditions are met:

• Abnormal wear
• Powdered fiber between strands
• Broken or cut fibers
• Variations in the size or roundness of strands
• Discoloration or rotting
• Distortion of hardware in the sling
• Acid or caustic burns
• Melting or charring of any part of the sling surface
• Snags, punctures, tears, or cuts
• Broken or worn stitches or
• Distortion of fittings

Figure 14.12 provides examples of synthetic sling criteria that would warrant removal from service.

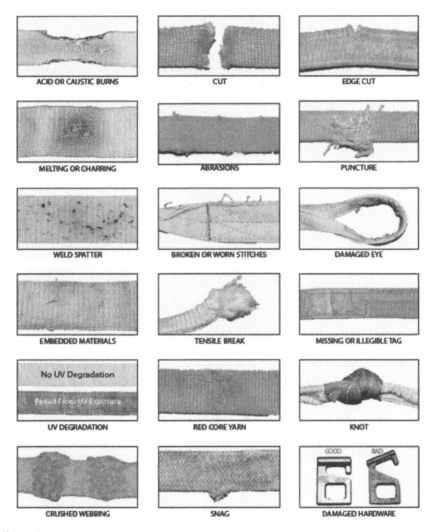

Figure 14.12

REVIEW QUESTIONS

1. Compare and contrast an overhead bridge crane with a gantry crane. Describe these components of an overhead crane: runway rails, bridge, trolley, and hoist.
2. What are the training/credential requirements for an overhead crane operator?
3. What is a crane "frequent" inspection? What must be examined as part of this inspection type? Who usually conducts this type of inspection?

4. What crane/hoist components must receive a documented, monthly inspection?
5. What is a crane "periodic" inspection? What must be examined as part of this inspection type? Who usually conducts this type of inspection?
6. Name five rules for the safe handling of a crane load.
7. What are slings? What is rigging? How do sling configurations affect the capacity of rigging?
8. What must be attached to all types of rigging?
9. What criteria would require alloy chain slings to be removed from service? Wire rope? Synthetic straps?

REFERENCES

Occupational Safety & Health Administration [OSHA]. (2016). Regulations (Standards-29 CFR 1910.179). Retrieved from https://www.osha.gov/laws-regs/regulations/standardnumber/1910/1910.179
Occupational Safety & Health Administration [OSHA]. (2019). Regulations (Standards-29 CFR 1910.184). Retrieved from https://www.osha.gov/laws-regs/regulations/standardnumber/1910/1910.184

APPENDIX A—OSHA TABLE N-184-1—ALLOWABLE CHAIN SIZE

Table N-184-1-Minimum Allowable Chain Size At Any Point of Link

Chain size, inches	Minimum allowable chain size, inches
1/4	13/64
3/8	19/64
1/2	25/64
5/8	31/64
3/4	19/32
7/8	4 5/64
1	13/16
1 1/8	29/32
1 1/4	1
1 3/8	1 3/32
1 1/2	1 3/16
1 3/4	1 13/32

Chapter 15

Machinery and Machine Guarding

INTRODUCTION AND SCOPE

Machines are essential components in all industrial environments and production facilities. Industrial machines are powerful and unforgiving. The moving parts and components of machines can easily pinch or cut through parts of the body. Failure to protect employees from the hazards of machinery can result in serious injuries and amputations.

Most or all machines have a power source (e.g., engine). The energy from the power source is transmitted, via a power transmission (e.g., belts and pulleys, gears, shafts), to the point of operation where the machine performs its action (e.g., cutting, bending, shearing, grinding) to achieve its goal. Machines are controlled by operators at operating panels or stations. The hazardous movement and actions of machines must be guarded to protect operators and nearby employees.

Machines can pose a variety of hazards related to their energy source, power transmission, and point of operation. Any machine hazard capable of posing serious harm to employees must be guarded or protected. Common machine guards include fixed guards or enclosures, interlocked guards, electronic safety devices, self-adjusting guards, and adjustable guards. These types of machine guards and protective devices have their different advantages and disadvantages based on their application.

In general, the point of operation and power transmission, along with any other hazardous moving parts of any machine, must be guarded. OSHA provides specific standards for the guarding of woodworking machines, abrasive wheel machinery, some specific machinery, and power transmission. This chapter focuses on guarding methods and requirements applicable to most machinery, as well as requirements specific to woodworking and abrasive wheel machinery. Requirements for specific machinery (e.g., power presses and forging machines) are omitted from this chapter; however, if you work for an employer who uses these types of machines, it is important to be familiar with the standards regulating them.

Hand and power tools are used across all industries. Hand and power tools must be used as designed and maintained in safe, working order. Power tools require specific types of switches depending on their use. OSHA provides general requirements for power tools as well as requirements for specific tools. In addition to specific power tool requirements, many of OSHA's electrical safety standards will apply to electrically powered power tools such as requirements for electrical grounding, insulation, power cord condition, and others. This chapter gives a general overview of OSHA provisions for tools and requirements for power tool switches, portable circular saws, portable grinders, and explosive actuated tools.

The OSHA standards covered in this chapter include:

- 29 CFR 1910 Subpart O—Machinery and Machine Guarding:
  - 29 CFR 1910.212—General requirements for all machines
  - 29 CFR 1910.213—Woodworking machinery requirements
  - 29 CFR 1910.215—Abrasive wheel machinery
  - 29 CFR 1910.219—Mechanical power transmission apparatus
- 29 CFR 1910 Subpart P—Hand and Power Tools
  - 29 CFR 1910.242—Hand and portable powered tools and equipment, general
  - 29 CFR 1910.243—Guarding of portable powered tools
  - 29 CFR 1910.244—Other portable tools and equipment

The OSHA standards not covered in this chapter include:

- Under CFR 1910 Subpart O—Machinery and Machine Guarding:
  - 29 CFR 1910.214—Cooperage machinery
  - 29 CFR 1910.216—Mills and calenders in the rubber and plastic industries
  - 29 CFR 1910.217—Mechanical power presses
  - 29 CFR 1910.218—Forging machines

NARRATIVE

Carri is a high school senior, a good student, and the pitcher of the softball team. Carri received a full scholarship to play softball at a university well known for its softball program. She is excited that she gets to go to her favorite school for free to play the sport she loves and study her field of interest. Carri is passionate about pursuing a career in Marine Biology. Not only is the school known for its success in softball, it is well known for its Marine Biology program. Carri is eager for college. She has all intentions of making an impact for her college team by throwing strikes with her right arm, and she is excited to have a career working with dolphins and aquatic life.

Between classes to finish her senior year, Carri works at a bakery. The bakery is not just a kitchen, but a large industrial facility that makes packaged baked goods. This facility has a large production kitchen with industrial baking equipment and a production line where the baked goods are processed and packaged. Carri works in the Prep. Department spending most of her shift in the kitchen. Carri adds large amounts of ingredients in huge mixers. These mixers are much larger and more powerful than household kitchen mixers. Carri was impressed by the scale of the equipment when she started working at the facility but has gotten used to the large equipment and machinery. These large, industrial mixes are old models. They are not equipped with adequate guarding at the point of operation. Typically, large mixes have a guard that must be closed in order to operate. This guard is called an interlocked guard. When the guard is locked in place, the electric circuit is closed, and it allows the mixer to turn on. If the guard is removed, the machine shuts off. This way, it is physically impossible for employees to stick their hands in the bowl while the mixer is running. The mixers Carri uses do not have an interlocked guard. The lack of proper machine guarding cost Carri her future.

Carri has made hundreds if not thousands of batches of the pastry she is preparing today. Carri adds dry and wet ingredients to the very large mixing bowl. She then turns the power on to the mixer. The dough hooks quickly and forcefully kneads and mixes the ingredients into dough. Carri is on her first batch of the day. Employees often manipulate the dough with a long wooden utensil as the dough hook mixes and kneads. Today, Carri's wooden spoon is not where it usually is. Another employee must have walked away with it. Without the spoon, Carri reaches in the bowl to use her hand to try to quickly push ingredients toward the center of the bowl while trying to keep her hand out of harm's way. Carri reaches in to the bowl and miscalculates her movement. Her right hand is caught in the powerful, unforgiving dough hook. This pulls her entire right arm into the rotating machinery. Carri screams in agony as bones in her right wrist, arm, and elbow are broken in multiple places. Luckily, Carri is coordinated enough to hit the off-button with her free, left hand before she is pulled in to her shoulder. The damage to her right arm is substantial.

Carri is rushed to the emergency room where she undergoes emergency surgery. Doctors informed her the damage to her bones, ligaments, and tendons is extensive and she likely will never have her arm back to normal even with multiple surgeries and extensive therapy. Carri is right-handed, so her softball career ends with a debilitating injury. Carri loses her scholarship, because she is unable to practice and play with the team. Carri can no longer afford to go to the same university. Carri reluctantly decides to attend community college instead. She loses out on the chance to major in Marine Biology and attends as undecided. Since her injury, Carri underperforms in school. Carri suffers from depression and anxiety from the incident. She spends most of her time in class daydreaming about the life she could have had if it wasn't for her horrific injury.

SUMMARY OF OSHA STANDARDS

Machine Guarding Introduction and General Requirements

Machine guarding methods must be used to protect machine operators and employees from hazards such as the point of operation of a machine, in-running nip points, rotating parts, flying debris, and sparks. Examples of machine guarding methods include fixed barrier guards, self-adjusting guards, interlocked guards, and electronic safety devices. In general, any machine part, function, or process that might cause serious injury has to be guarded. It is important to recognize there is no law or requirement for machines to be manufactured and sold with OSHA-compliant machine guards. Just because a machine is purchased (or was purchased a long time ago) and has unguarded moving parts does not mean it is excused from complying with OSHA standards. Machine guards shall be affixed to the machine and secured. Machine guards themselves shall not create a hazard. Related to machine guarding, employers should enforce policies that prevent machine operators from wearing long hair, loose clothing, or dangling jewelry. These items create a hazard that can be caught in and pull the employee into rotating or moving parts. Before reading forward, it is helpful to identify machine operational areas, common types of machine hazards, and common types of machine guards.

Machine Operational Areas

In general, a machine can be separated into three primary operational areas that are important for machine guarding. The operating controls of a machine are located at the operator's station or the operating panel. In figure 15.1, the operating controls of the shearing machine are located to the right of the machine on the gray operating panel. These are the buttons, knobs, switches, and so on that activate and control the machine. The point of operation of a machine is where the machine does its job. In figure 15.1, the point of operation of the shear is protected by a fixed fence guard. This is where the shear in

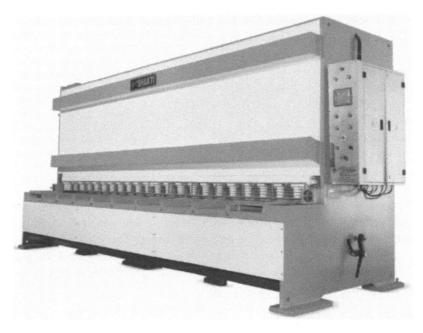

Figure 15.1

figure 15.1 moves downward to cut metal. The point of operation is where a machine cuts, shears, bends, grinds, shapes, and so on. The point of operation is where the material is machined and the machine's action is performed. The point of operation of a machine whose operation exposes an employee to an injury shall be guarded and shall be so designed to prevent the operator from having any part of his or her body in the danger zone during the operating cycle. In figure 15.2, the point of operation of the planer is where the blade

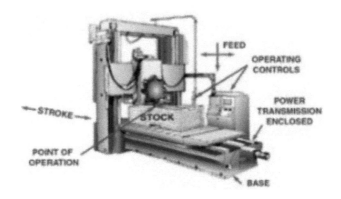

Figure 15.2

cuts the stock piece of wood. The following are some of the machines which usually require a point of operation guard:

- Guillotine cutters
- Shearing
- Power presses
- Milling machines
- Planters and jointers
- Portable power tools
- Blades

The power transmission is the area of the machine that transmits the power source's energy to the point of operation. In figure 15.3, an unguarded power transmission made up of belts and pulleys is shown. The power transmission of a machine can be belts and pulleys, chains, gears and sprockets, shafts, and other components or a system of components that moves energy from the source (engine) to the point of operation.

Common Machine Hazards

Basic types of machine hazards can be categorized as hazardous motions and hazardous actions. Common hazardous motions include rotating (including

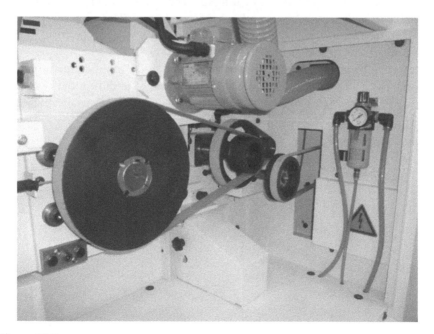

Figure 15.3

in-running nip points), reciprocating, and transversing. Common hazardous actions include cutting, punching, shearing, and bending. In addition to guarding hazardous motions and actions, machines designed for a fixed location shall be securely anchored to prevent walking or moving.

Rotating Motion and In-Running Nip Points

Rotating parts can grip clothing, cause harm from skin contact, and force limbs into dangerous positions. Examples of rotating parts include flywheels, shafts, gears, cams, and collars. Danger increases when rotating parts have projections like bolts, screws, or other parts. Figure 15.4 depicts rotating hazards.

In-running nip points are caused by rotating machinery. In-running nip points can be caused by parts rotating in the opposite direction while next to each other (e.g., intermeshing gears). This is shown in figure 15.5. Nip points can also be created between rotating parts and tangentially moving parts (e.g., belt running on pulley, chain running on sprocket). This is shown in figure 15.6. In-running nip points can also be created between rotating and fixed parts (e.g., flywheels, abrasive grinding wheels' point of operation). This is shown in figure 15.7. Clothing and body parts can be pulled into in-running nip points

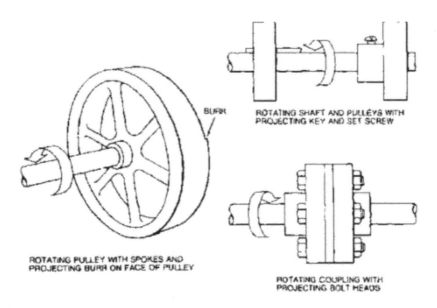

Figure 15.4

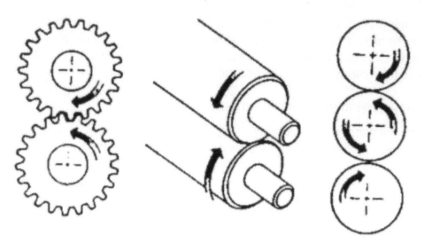

Figure 15.5

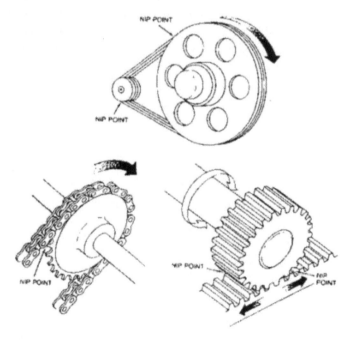

Figure 15.6

Reciprocating and Transverse Motion

Reciprocating motion can be hazardous due to up and down or back and forth motion. A worker can be struck by or caught between machinery and fixed

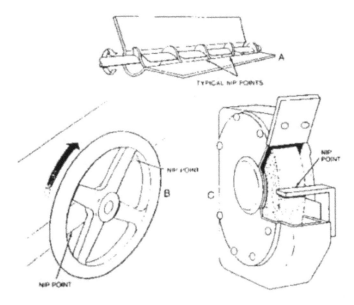

Figure 15.7

objects. Figure 15.8 depicts a hazard of a reciprocating table. Transversing motion (straight, continuous line) can be a hazard because a worker can be caught in a pinch or shear point by the moving part. Figure 15.9 depicts a transverse motion hazard.

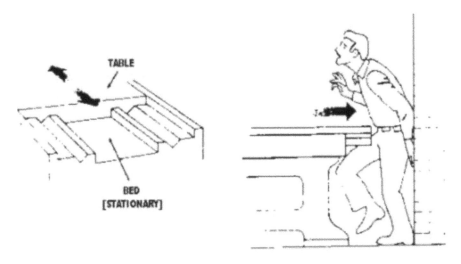

Figure 15.8

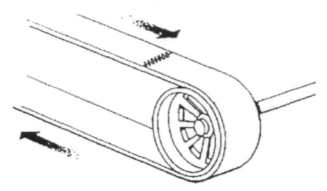

Figure 15.9

Machine Actions (Cutting, Punching, Shearing, and Bending)

Actions of a machine can be hazardous such as a cutting action. Cutting action may involve rotating, reciprocating, or transverse motion of a blade. Cutting action presents a point of operation hazard where fingers or body parts can be cut. Scrap material, debris, chips, and sparks can also be thrown from the point of operation. Common examples of machines that involve cutting hazards include band saws, table saws, drilling machines, turning machines (lathes), and milling machines. Figure 15.10 illustrates cutting hazards.

Punching action results when power is applied to a slide or ram for stamping or punching (cutting out) metal material. The primary hazard of this action is at the point of operation where metal is punched out. Figure 15.11 depicts a punching action.

Shearing action involves applying power to a shear or knife to trim or cut metal. The primary hazard of this action is at the point of operation where the shear cuts metal. Figure 15.12 shows a shearing action.

Bending action results when power is applied to a slide or ram to bend metal (rather than punch through it). The primary hazard of this action is at the point of operation and is similar to a punching action. Figure 15.13 shows a bending action.

Common Machine Guards and Devices

The most common types of machine guards are fixed or enclosed guards, interlocked guards, self-adjusting guards, and adjustable guards. These guards have their advantages and disadvantages. Some are better than others, but not all guards are feasible for all applications. In addition to mechanical guards, electric safety devices can be used for employee protection. This section describes common applications for machine guards and safety devices and gives examples of each along with their advantages and disadvantages.

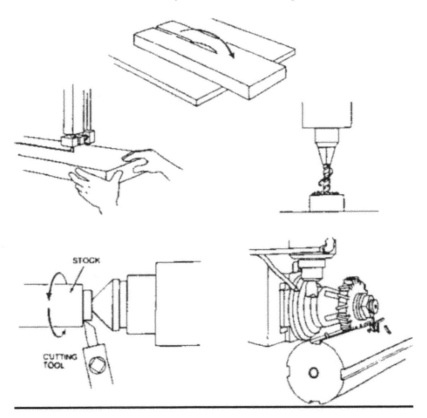

STOCK

CUTTING
TOOL

Figure 15.10

Fixed Guards or Enclosures

A fixed guard or enclosure is installed somewhat permanently. Usually, these guards are only removed with necessary tools if maintenance is required. If a fixed guard or enclosure is removed, the control of hazardous energy (lockout/tagout) is required. These guards theoretically provide the most protection, but cannot be used in all types of applications. They are most commonly used for guarding power transmission hazards, since the only time power transmission would be accessed is for maintenance and adjustment.

- Common Application:
  - Power transmission enclosures (e.g., enclosed belt and pulleys) (figure 15.14)
  - Fixed guards with restricted openings for the point of operation (e.g., fence guard of shearing machine) (figure 15.15)

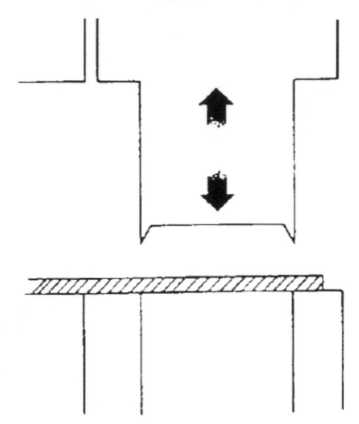

Figure 15.11

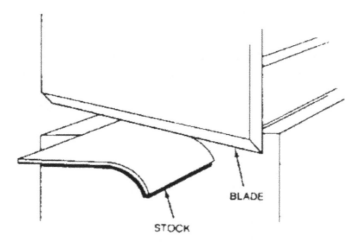

Figure 15.12

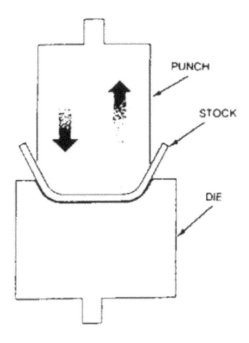

PUNCH

STOCK

DIE

Figure 15.13

Figure 15.14

Figure 15.15

- Advantages:
  - Can be constructed (custom fabricated) to suit many applications
  - Can provide maximum protection, do not rely on adjustment
  - Minimal maintenance
  - Durable
- Disadvantages:
  - May interfere with visibility or otherwise may get in the way
  - Limited to specific operations where required access is limited
  - Machine adjustment or repair may require removal of the guard (and the control of hazardous energy)

Fixed guards or enclosures with openings (as seen in figures 15.14 and 15.15) shall be designed such that openings are small enough to not allow fingers or body parts to reach the hazard. OSHA gives specific standard requirements for guards used to protect the point of operation of mechanical power presses; however, this guidance can be applied to other hazards that are protected by guards with openings. The table in appendix A of this chapter shows distances guards with openings shall be positioned from the danger zone in accordance with the size of the openings.

Interlocked Guards

Interlocked guards are commonly used to guard the point of operation of machines. Interlocked guards are electrically connected to the machine. If the interlocked guard is opened, an electric switch is opened, and this shuts down the equipment. When an interlocked guard is opened, hazardous motion is stopped, and the machine is placed in a safe state. There are some major advantages to using interlocked guards for employee protection.

- Common Application:
  - Point of operation of machines (e.g., enclosed lathe) (figure 15.16)
- Advantages:
  - Can provide maximum protection (shuts off the machine when opened)
  - Allows temporary control of energy for routine production tasks
  - Do not have to remove a fixed guard to access
- Disadvantages:
  - Can be defeated or can fail
  - Requires maintenance
  - Cannot be used in place of lockout/tagout for maintenance tasks

Electronic Safety Devices (Presence Sensing Devices)

Similar to interlock guards, electronic safety devices or presence sensing devices are connected to the machine's electrical system. When the device detects the presence of a worker's hand or body, the device opens the electrical circuit and stops the machine's hazardous motion. Two common examples of presence sensing devices are light curtains and presence sensing mats. When the beams of a light curtain are broken, hazardous motion of the machine is stopped. Conversely, presence sensing devices can be designed to only allow the machine to run when the operator is standing in a safe location (e.g., standing on a presence sensing mat at the operator panel and not in a danger zone). These

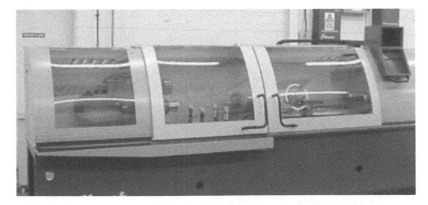

Figure 15.16

Figure 15.17

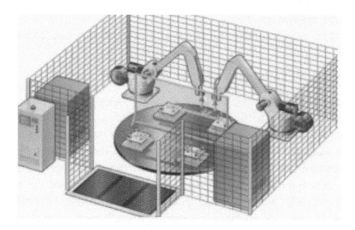

Figure 15.18

devices have some real advantages for protection from hazardous motion; however, they must be maintained, adjusted, and calibrated to ensure effectiveness.

- Common Application:
    - Point of operation of machines (e.g., light curtain for press) (figure
      15.17)
    - Robotic enclosures (e.g., presence sensing mat in robot enclosure) (figure
      15.18)
- Advantages:
    - Can allow more free motion of the operator
    - Simple to use and can be effective
    - Allows temporary control of energy for routine production tasks
    - Do not have to remove a fixed guard to access
    - Can be designed for presence to allow operation or presence to cease operator

- Disadvantages:
  - Can fail
  - Requires alignment, calibration, and maintenance to ensure effectiveness
  - Not appropriate from some environmental conditions that can impair sensors or effectiveness

OSHA provides some specific guidance for the distance of a presence sensing device's sensing field (e.g., light curtain) from the point of operation. This guidance is specifically geared toward mechanical power presses, but the guidance should be applied to any use of such devices to protect the point of operation of the machine. The safe distance (D) from the sensing field to the point of operation shall be no greater than the distance determined using equation 15.1, where T is the stopping time of the machine and 63 inches/second is the typical hand speed.

$$D = 63 \frac{\text{inches}}{\text{second}} \times T$$

(15.1)

Self-Adjusting Guards

Self-adjusting guards can accommodate stock of different sizes. They provide a barrier that moves according to the size of the material being machined. Protection is not always complete, since the guard must move.

- Common Application:
  - Point of operation of machines (e.g., table saw) (figure 15.19)
- Advantages:
  - Accommodates different stock sizes
  - Provides protection for part of the machine not engaged in the stock
- Disadvantages:
  - Protection is not always complete since the guard moves based on the size of stock
  - Operator must ensure guard is working properly
  - May required adjustment and frequent maintenance
  - Guard can be defeated easily by the operator
  - May interfere with visibility

Adjustable Guards

Adjustable guards must be adjusted by the operator to protect the point of operation as much as possible. These guards depend on the operator using

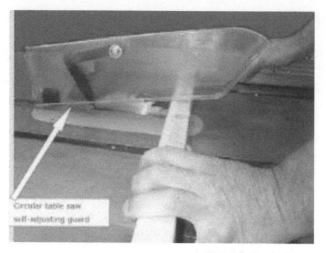

Circular table saw
self-adjusting guard

Figure 15.19

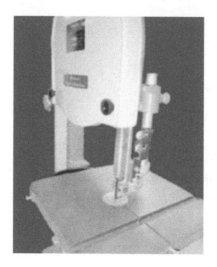

Figure 15.20

correctly and do not always provide complete protection, so for that reason, they are the weakest type of guard.

- Common Application:
  - Point of operation of machines (e.g., band saw) (figure 15.20)
- Advantages:
  - Can accommodate different stock sizes
- Disadvantages:
  - Depends on the operator to adjust properly

- ○ Hands can still enter the danger zone
- ○ Protection is not always complete
- ○ Requires frequent adjusting and maintenance
- ○ Guard can be defeated by the operator
- ○ May interfere with visibility

Common Woodworking Machinery Requirements (Saws)

Machine Controls and Equipment

The operator of woodworking machines must be able to turn off power to the machine without leaving his or her position at the point of operation. Provisions shall be made to prevent any woodworking machine from automatically restarting upon any restoring of power after a power outage. All power controls of woodworking machines shall be located within easy reach of the operator while at the operator's regular work location. The operator shall not be required to reach over the cutter to make adjustments. If woodworking machines are operated by foot treadles, the treadle shall be guarded against accidental activation.

Hand-Fed Ripsaws and Crosscut Table Saws

A hand-fed ripsaw is a circular saw that requires wood to be fed into the blade by hand. Figure 15.21 depicts a hand-fed ripsaw. A ripsaw is a saw that cuts wood parallel to the direction of the grain. This is called a rip cut.

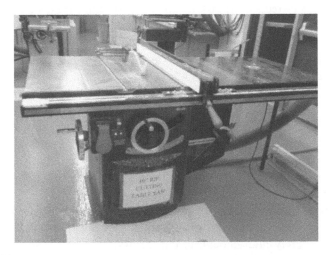

Figure 15.21

Figure 15.22

Figure 15.22 shows the difference between a rip cut and crosscut. Each hand-fed ripsaw (e.g., typical table saw used for rip cuts) shall be guarded by a hood (self-adjusting guard) that enclosed the portion of the saw blade above the table and the portion of the blade above the material being cut (as seen in figure 15.19). The self-adjusting hood-guard shall be durable and protect from flying debris.

In addition to a self-adjusting guard, all hand-fed ripsaws shall have a spreader to prevent material from squeezing the saw blade or being thrown back at the operator. The spreader (or riving knife) splits wood apart as it is being cut to prevent the squeezing of the blade. This prevents the kickback hazard (wood being thrown toward the operator). These saws used for rip cuts also are required to have non-kickback fingers located so as to oppose the trust of potential kickback of material. Both spreaders and non-kickback fingers work together to reduce the risk of kickback during rip cuts and are seen in figure 15.23. Kickback is much more likely during rip cuts as compared to crosscuts.

A crosscut table saw is a table saw used for crosscutting wood. Crosscut tables often use a crosscut sled (a guide or device that pushes wood through the blade). All crosscutting table saws require a self-adjusting guard like rip cut table saws. The only difference in requirements for table saws used for only crosscutting is crosscut saws do not require a spreader nor non-kickback fingers. It is important that only ripsaws (those equipped with kickback protection) are used for rip cuts. Crosscut saws (those without kickback protection) shall only be used for crosscuts. Rip cutting table saws can be used for either rip cuts or crosscuts.

There is a safe work practice worth mentioning for table saws used in combinations with guards. Table saw blades should be adjusted to be as low as possible (barely high enough to cut through stock thickness). This way,

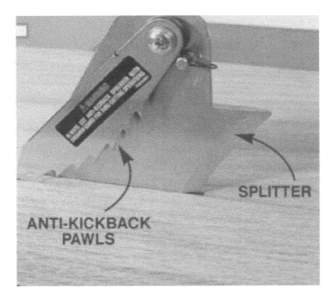

SPLITTER

ANTI-KICKBACK
PAWLS

Figure 15.23

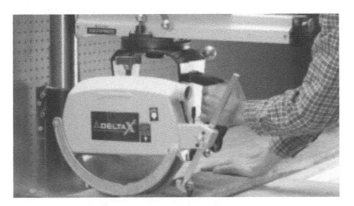

Figure 15.24

the majority of the point of operation is within the work piece. This reduces severity of injury in the event of incidental contact during cutting.

Radial Arm Saws

A radial arm saw consists of a blade that can be moved in and out for rip cutting or crosscutting. Figure 15.24 shows a radial arm saw. The upper portion of the blade must be completely enclosed (from the top of blade to the center point). The lower portion of the blade shall be guarded by a self-adjusting guard to give maximum protection as possible based on the size of the stock being cut. If used for rip cutting, the saw must be equipped with non-kickback fingers. Radial arm saws must have an adjustable stop capable of preventing forward travel based on the position necessary to complete the cut. The saw must automatically return to its starting position.

Band Saws

Band saws have a blade that is a continuous band. Figure 15.25 shows a vertical band saw and figure 15.26 shows a horizontal band saw. For vertical band saws, material is pushed into the blade. For horizontal band saws, the blade is lowered through the material. All portions of the continuous blade must be enclosed except for the portion required to make the cut based on the size of the material. The point of operation is typically protected by a manually adjusted guard. Band saw wheels on which the blade runs shall be fully enclosed.

Inspection and Maintenance of Woodworking Machinery

Damaged, dull, or improperly tensioned saws shall be immediately removed from service. Saws to which gum or gunk has adhered to the sides shall be cleaned. Arbors (on which blades sit) shall be free of play or wiggle. A saw

Figure 15.25

Figure 15.26

motor drives the arbor which causes a blade to rotate. Woodworking machinery shall be cleaned effectively to ensure guards work properly and to reduce fire hazards. Any cracked saw blades shall be removed from service. When hands or fingers could encroach the danger zone to feed material into a saw blade, push sticks or push blocks shall be provided. Figure 15.27 depicts a push stick being used.

Abrasive Wheel Machinery

Abrasive wheel machines include pedestal and bench grinders. Figure 15.28 depicts an example of a bench grinder. A pedestal grinder is similar, but it sits on top of a pedestal. Abrasive wheels shall be guarded such that the spindle

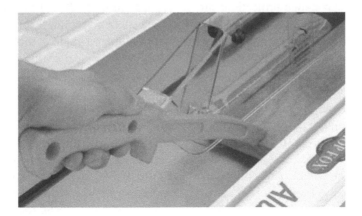

Figure 15.27

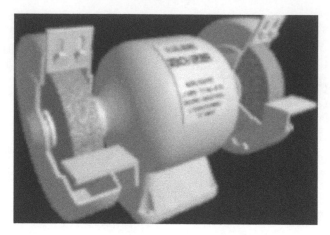

Figure 15.28

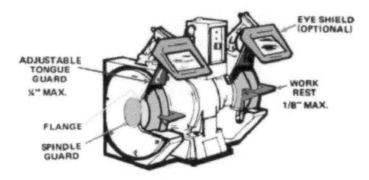

Figure 15.29

end, nut, and flange projections are covered. Figure 15.29 will help identify components that will be referred to throughout this section. The work rest on a grinder is the component used to support the workpiece that is being grinded. Work rests shall be adjusted closely to the wheel with a maximum opening of a one-eighth inch between the wheel and the work rest. This prevents the workpiece from being jammed in the in-running nip point between the wheel and the work rest. Bench and pedestal grinders shall have a minimum angular exposure that shall not exceed 90 degrees of the grinding wheel (no more than one-fourth of the total grinding wheel can be exposed). This exposure shall begin at a point not more than 65 degrees above the horizontal plane of the wheel. This is illustrated in figure 15.30. The tongue guard shall be adjusted to be no more than one-fourth inch from the wheel. The tongue guard helps contain shards of the wheel if it is damaged during rotation.

Any and all abrasive wheels must be inspected before mounting. In doing so, the wheel must be subject to a ring test to make sure it is free from cracks.

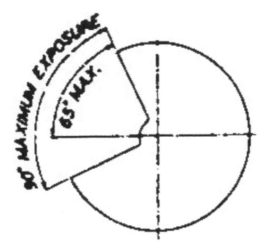

Figure 15.30

To conduct a ring test, the wheel must be tapped gently with a light non-metallic tool. If a "dead" sound is heard, the wheel shall not be used. When tapped the wheel shall make a "ringing" sound to indicate it is free from cracks or damage. Wheels must be dry and free from dust when conducting a ring test. The speed of the grinder shall be examined to ensure it does not exceed the maximum operating speed marked on the wheel.

Mechanical Power Transmission Apparatus

Mechanical power transmission apparatus (e.g., belts) shall be guarded unless they meet specific exemptions. These exemptions depend on the type of belt used and size. These exemptions are not covered in this chapter. In general, all moving parts of the power transmission apparatus should be guarded if they pose a serious risk of harm. Horizontal belts and pulleys do not need to be guarded if they are more than 7 feet above the floor or walking platform (though, they should be guarded if incidental contact is possible). Gears, sprockets, and chains shall be guarded by an enclosure. Those gears that are hand-operated are not required to be guarded (e.g., drive train of a hand-powered garage door opener) unless they otherwise pose a serious risk of harm.

Tools

General Requirements

The employer is responsible for the safe condition of all tools and equipment used by employees. This includes both manual, hand tools and power tools. This includes any tools and equipment employees may bring from home.

Hand tools must be used safely, maintained free of defects or damage, and used only as designed.

Compressed air used for cleaning (e.g., blowing the dust off work stations, body, or clothes) must be reduced to less than 30 pounds per square inch (psi). When compressed air is used for cleaning, employees must be wearing appropriate personal protective equipment (e.g., safety glasses).

Switches and Controls

The operating control on handheld power tools shall be located to prevent accidental operation.

The following power tools must be equipped with a "constant pressure switch" such that, when pressure is released from the button/control, power is shut off:

- Portable circular saws with a diameter greater than 2 inches
- Portable chain saws

The following power tools must be equipped with a "constant pressure switch" and may have a "lock-on" control, provided that turnoff can be accomplished by a single motion of the same finger that turns it on:

- Portable power drills, tappers, and fastener drivers
- Horizontal, vertical, and angle grinders with wheels greater than 2 inches in diameter
- Disc sanders with discs greater than 2 inches diameter
- Belt sanders
- Reciprocating saws
- Scroll and jigsaws with blade width being greater than one-fourth inch
- Tools similar to these listed

The following power tools may be equipped with a simple "on-off" control:

- Grinder with wheels 2 inches in diameter or less
- Disc sanders with discs 2 inches in diameter or less
- Portable routers, planers, nibblers, shears
- Scroll saws and jig saws with blade width being one-fourth inch or less
- Other power tools similar to those listed here and that do not require a constant pressure switch

Basic Electrical Requirements

There are a few key electrical requirements for electrically powered portable power tools. Electrical tools shall be grounded or double insulated. This means

Figure 15.31

the electrical cord plug shall have a grounding prong. If it does not, the tool shall be double insulated and have a visible marking indicating it as such. The double-insulated marking is a square inside of a square seen in figure 15.31. Insulation of electrical power tool cords shall not be damaged or frayed. If equipped with a grounded prong, the prong shall not be damaged or removed.

Requirements for Portable Circular Saws

All power-drive, portable circular saws having a blade greater than 2 inches in diameter shall be equipped with a self-adjusting guard. The guard shall always cover the top of the blade. The lower portion of the guard (below the base plate) shall cover the saw to the depth of the teeth except for the minimum arc required to allow proper retraction and contact with the workpiece. When not in use, the guard shall automatically and instantly return to the covering position. A circular saw with a self-adjusting guard is seen in figure 15.32.

Requirements for Portable Abrasive Wheels (Grinders)

OSHA provides requirements for portable grinders with discs larger than 2 inches. Most commonly, portable abrasive wheels are vertical or angle grinders. A safety guard must cover the spindle end, nut, and flange projections of the grinder. There are very few exceptions for the requirement of this guard. Guards on these grinders shall have a maximum exposure angle of 180 degrees. The guard shall be located between the operator and the wheel during use. Figure 15.33 depicts the proper guard in place. The handle seen in figure 15.33 is not required by OSHA.

Figure 15.32

Figure 15.33

Before mounting a grinding wheel on a portable grinder, the wheel must be inspected to make sure it is not damaged. The maximum operating speed of the machine should be examined to be certain it does not exceed the maximum operating speed marked on the wheel. Also, operators shall conduct a ring test on grinding wheels as described previously in this chapter for abrasive wheel machinery

REVIEW QUESTIONS

1. What machine hazards are required to be guarded? What are the operational areas of a machine?
2. Name and describe the types of hazardous machine in motion and action.

3. Compare and contrast common types of machine guards and electronic protective devices.
4. What are the required machine guards and protective devices for a table saw used for rip cutting?
5. How far must a tongue guard and a tool rest be from the wheel of a grinder?
6. What is a ring test? How do you perform a ring test? What is the purpose?
7. Based on the narrative, what type of guard should the mixer likely have been fitted for?

REFERENCES

https://www.osha.gov/Publications/Mach_SafeGuard/chapt1.html

Occupational Safety & Health Administration [OSHA]. (1984). Regulations (Standards-29 CFR 1910.212). Retrieved from https://www.osha.gov/laws-regs/regulations/standardnumber/1910/1910.212

Occupational Safety & Health Administration [OSHA]. (1984). Regulations (Standards-29 CFR 1910.213). Retrieved from https://www.osha.gov/laws-regs/regulations/standardnumber/1910/1910.213

Occupational Safety & Health Administration [OSHA]. (1996). Regulations (Standards-29 CFR 1910.215). Retrieved from https://www.osha.gov/laws-regs/regulations/standardnumber/1910/1910.215

Occupational Safety & Health Administration [OSHA]. (2013). Regulations (Standards-29 CFR 1910.2115). Retrieved from https://www.osha.gov/laws-regs/regulations/standardnumber/1910/1910.217

Occupational Safety & Health Administration [OSHA]. (2004). Regulations (Standards-29 CFR 1910.219). Retrieved from https://www.osha.gov/laws-regs/regulations/standardnumber/1910/1910.219

Occupational Safety & Health Administration [OSHA]. (1978). Regulations (Standards-29 CFR 1910.241). Retrieved from https://www.osha.gov/laws-regs/regulations/standardnumber/1910/1910.241

Occupational Safety & Health Administration [OSHA]. Regulations (Standards-29 CFR 1910.242). Retrieved from https://www.osha.gov/laws-regs/regulations/standardnumber/1910/1910.242

Occupational Safety & Health Administration [OSHA]. (2007). Regulations (Standards-29 CFR 1910.243). Retrieved from https://www.osha.gov/laws-regs/regulations/standardnumber/1910/1910.243

Occupational Safety & Health Administration [OSHA]. (1984). Regulations (Standards-29 CFR 1910.244). Retrieved from https://www.osha.gov/laws-regs/regulations/standardnumber/1910/1910.244

APPENDIX A—OSHA TABLE O-10 DISTANCES FROM POINT OF OPERATION BASED ON GUARD OPENINGS

Distance of opening from point of operation hazard	Maximum width of opening
1/2 to 1 1/2	1/4
1 1/2 to 2 1/2	3/8
2 1/2 to 3 1/2	1/2
3 1/2 to 5 1/2	5/8
5 1/2 to 6 1/2	3/4
6 1/2 to 7 1/2	7/8
7 1/2 to 12 1/2	1 1/4
12 1/2 to 15 1/2	1 1/2
15 1/2 to 17 1/2	1 7/8
17 1/2 to 31 1/2	2 1/8

Chapter 16

Welding, Cutting, and Brazing

INTRODUCTION AND SCOPE

Welding, cutting, and brazing (hot work) are common industrial operations. Welding or brazing involves creating a very high localized temperature to melt a filler material and heat surrounding metal to adjoin metal together. The two most common types of welding are electric arc welding and oxygen/fuel gas welding. Electric arcs and oxygen/fuel gas torches can also be used to cut through metal. Hot work can pose a variety of physical and health hazards such as metal fumes, toxic gases, harmful radiation, electricity, and fire. The hazards of hot work must be adequately controlled with a hot work permit procedure, adequate ventilation, proper PPE, proper equipment storage and maintenance, and other control methods as required by the OSHA standard. Figure 16.1 depicts a welding operation using an electric arc welder and electrode as the filler material (stick welding).

The two most common types of hot work are electric arc welding and oxygen/fuel gas welding. Oxygen/fuel gas welding (or cutting) combines oxygen with a flammable gas to weld metal together or to cut through metal. There are particular requirements for the safe use and storage of oxygen/fuel gas cylinders that commonly apply across industries. Uncommon requirements related to oxygen/fuel gas welding or cutting are omitted from this chapter. Figure 16.2 illustrates an oxygen/fuel gas welding or cutting system. Electric arc welding (or cutting) uses electricity to weld metal together or cut through metal. Electric arc welders must be used and maintained per the manufacturer's instructions, and precautions must be taken to avoid electrical shock. Requirements related to the design, installation, and maintenance of electric arc welding machines are omitted from this chapter. The primary

Figure 16.1

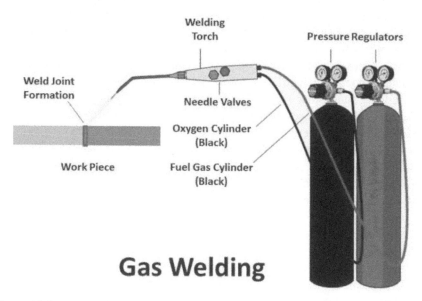

Figure 16.2

OSHA-required precautions covered in this chapter are related to fire protection and personnel protection that apply across both types of hot work.

The OSHA standards covered in this chapter include:

- 29 CFR 1910 Subpart Q—Welding, Cutting, and Brazing
  - ○ 29 CFR 1910.252—General requirements
  - ○ 29 CFR 1910.253—Oxygen fuel-gas welding and cutting
  - ○ 29 CFR 1910.254—Arc welding and cutting

Not included:

- Under 29 CFR 1910 Subpart Q—Welding, Cutting, and Brazing
  - ○ Additional considerations for hazard communication in welding, cutting, and brazing
  - ○ Requirements related to manifolding of gas/fuel gas cylinders, portable outlet headers, service piping systems, piping protective equipment, and acetylene generators
  - ○ Requirements related to resistance welding
  - ○ Requirements related to specific industrial applications (e.g., transmission pipelines)
  - ○ Requirements related to the design, installation, and maintenance of electrical arc welding machines

NARRATIVE

Molly is a young woman who works as a welder. She is very bubbly and kind-hearted. Molly is known to always be smiling and to make other people smile. In vocational school, she learned about the basic hazards of welding and how to protect herself and others. After getting her welding certificate, Molly is very excited to be hired by the local plant as a welding apprentice. The plant makes heating and air conditioning equipment. When she is hired, her employer assures her that she will enjoy working at the company.

On her first day of work at the company, Molly learns quickly working in the real world is not like welding school. Things are fast pace and there is a lot of pressure to work quickly. Molly's coworkers are not kind and bubbly like her, so it makes it hard to enjoy going to work. Despite this, Molly still smiles and does her best to be friendly. Molly works in the welding shop. It is a small, cinder block room. It has a welding table in the middle and welding equipment surrounding the table. Molly asks if the company has a snorkel she can use for local exhaust ventilation to remove the fumes. The supervisor tells her not to worry, because she can leave the door open and there is enough air circulation. He also claims the metal she will be welding on is a harmless alloy. Molly knows enough that some metals can produce harmful fumes, but she trusts her supervisor's word.

Molly goes to work in the welding shop for decades, welding on the same type of metal for years. Molly is now a fifty-year-old mother to a son who is in college. As she tries to keep up with her son's busy lifestyle, she notices she has been having trouble breathing. This is just the start of things. Trouble breathing turns into a persistent cough and chest pains. Molly is in and out of the hospital consistently with infections like bronchitis and pneumonia. After some time, Molly's doctor gives her the news she is diagnosed with lung cancer. The cancer is quite aggressive and it likely will be terminal within a year. Through tears, Molly asks her doctor what could have caused this, and her doctor mentions hexavalent chromium. With some further investigation, it turns out Molly has been welding on metal that contains chromium for years. The fumes of exposure to hexavalent chromium based on her occupation, found in stainless steel, are a known carcinogen that can cause lung cancer. Molly is forced to retire from the plant. In her final living year, despite the reality that her terminal illness was brought on by her employer's failure to control the hazardous fumes of welding, she still smiles. She refuses to spend her final days living in bitter resentfulness for her employer. She chooses forgiveness and focuses on being her bubbly-self for her son and her family. Molly dies at the age of fifty-one, leaving her husband and son behind.

SUMMARY OF OSHA STANDARDS

General Requirement and Protections for Welding, Cutting, and Brazing

Basic Fire Prevention and Protection

OSHA provides some general requirements for welding, cutting, and brazing (hot work). OSHA also points employers to NFPA Standard 51B, 1962 (Standard for Fire Prevention in Use of Cutting and Welding Processes) for more detailed information. If the object to be welded or cut cannot be moved away from fire hazards, then fire hazards must be moved away from the welding area to prevent incidental ignition of those fire hazards. If fire hazards or the welding/cutting operation cannot be relocated, then fire hazards shall be guarded or protected from heat, sparks, and slag (hot metal). If the welding area is surrounded by combustion construction material (e.g., wooden floor, wall, or ceiling), then precautions must be taken to prevent ignition of the construction material. Floors shall be swept free of combustible dust and debris. Combustible materials (e.g., cardboard boxes, rags, flammable liquids) shall be at least 35 feet away from the hot work if not otherwise guarded from sparks and heat. If the floor, wall, or ceiling has cracks or openings, through which sparks might go through and cause a fire or injure others, then measures must be taken to prevent sparks from entering those cracks or openings.

During hot work operations, there must be suitable fire extinguishing equipment available for instant use (in the welding area). Such equipment may include a pail of water, buckets of sand, or a portable fire extinguisher of the appropriate type. If a portable fire extinguisher is provided, hot work personnel expected to use it must have fire extinguisher training annually. A fire watch must be present for hot work if certain conditions exist. A fire watch is someone who must monitor the hot work area for a fire and be ready to extinguish it or otherwise sound an alarm to evacuate. A fire watch must be trained in proper extinguishment. A fire watch is required if:

• There are appreciable amounts of combustible material within 35 feet of the hot work
• There are appreciable amounts of combustible material further than 35 feet of the hot work but the material is easily ignitable by sparks
• There are wall or floor openings within 35 feet of the hot work that are unprotected

A fire watch shall be maintained for at least a half hour after completion of the hot work.

HOT WORK PERMIT

Before hot work is permitted to occur, it must be authorized by use of a written permit. To be authorized, the hot work area shall first be inspected by the person responsible for authorizing the hot work. The person responsible for authorizing work shall set the precautions that must be followed (e.g., combustibles 35 feet away, fire watch present if needed, etc.) to proceed with the hot work and authorize work in the form of a written permit. An example of a hot work permit can be found in appendix A of this chapter. Hot work is prohibited under certain conditions. Hot work is not allowed in areas not authorized by employer management, in building that have sprinkler systems when sprinklers are not working, in the presence of an explosive atmosphere, and in areas near large quantities of readily ignitable materials.

When practical, combustible materials shall be relocated at least 35 feet away from the hot work. Ducts that might carry sparks from the hot work to combustible materials shall be shut down if near the hot work area. Based on fire potential in the plant or facility, management shall establish areas appropriate for welding and cutting and establish procedures for safe hot work via an authorization permit. Welders or those employees performing hot work and their supervisors shall be suitably trained in safe operations of equipment and execution of the hot work permit process.

WELDING AND CUTTING ON CONTAINERS AND IN CONFINED SPACES

Containers and vessels that store or are used to contain flammable materials shall be thoroughly emptied, cleaned, and purged to ensure no remaining vapors will be ignited or produce toxic vapors when welding. Purging means to flush out vapors with an inert (nontoxic, nonflammable) gas such as nitrogen. Similar processes need to occur when welding on pipes. Pipes shall be disconnected and blanked and bled. This means material in the pipe must be blocked from entering the section of the pipe to be worked on, and material must be drained out in that section of pipe.

When welding in confined spaces, ventilation must be adequate to prevent the accumulation of harmful gases or vapors, prevent oxygen deficiency, and prevent ignition of potentially flammable substances. When welding in confined spaces with a small opening, when accumulation of hazardous substances is possible, means shall be provided to quickly remove the welder in case of emergency. This is typically done by attaching a lifeline to the worker in a way so he or she can be pulled out without his or her body becoming jammed in the small exit opening. An attendant shall monitor the confined space during these operations, and pre-planned rescue procedures must be in place. When arc welding in a confined space for a substantial period of time, such as overnight or over a break, all electrodes and their holders shall be removed. If gas welding or cutting for a substantial period of time in a confined space, gas cylinders shall be shut off and torches and hoses removed from the confined space, during breaks or overnight, to prevent accidental accumulation of oxygen or gas inside the confined space.

GENERAL PERSONNEL SAFETY AND HEALTH DURING HOT WORK

Welding and cutting PPE shall be selected to adequately protect personnel. Welding helmets or hand shields shall be used during all arc welding or arc cutting operations. Helpers or attendants shall be provided proper eye protection as well. These helmets or shields shall protect the face, neck, and ears from direct radiant energy (ultraviolet light) from the arc. Figure 16.3

Figure 16.3

shows a typical welding helmet. Ultraviolet light (UV) radiation is especially dangerous to the unprotected eye. The lens of helmets or shields must have a marking that identifies the shade and level of protection from UV light. The proper type of shade (shade number) depends on the welding operation being conducted. The OSHA welding and cutting standard provides a table of shade recommendations. Goggles or other suitable eye protection shall be used during gas welding or oxygen cutting operations. Goggles must be ventilated to prevent fogging.

During electric arc welding, adjacent workers and those who walk by must be protected from UV light. This can be achieved using welding screens that filter out UV light or by providing employees with appropriate eye protection. Figure 16.4 shows the use of welding screens.

Employees exposed to the hazards created by hot work must be provided protective clothing depending on the type of operations. It is up to the employer to determine what protective clothing is necessary by hazard assessment. Fire-resistant jackets or sleeves are common. Generally, hot work attire and PPE include long pants, welding jacket, welding helmet/protective eyewear, and heat-resistant gloves. First-aid equipment shall be available at all times in facilities where hot work occurs.

Measures must be taken to prevent hazardous levels of exposure to hazardous fumes, gases, and dust involved with welding and cutting. Local exhaust ventilation or general ventilating systems shall be provided to keep the amount of toxic fumes, gases, or dust below the maximum allowable concentration as specified in OSHA 1910.1000. Whenever welding or cutting operations

Figure 16.4

involve metals or metal coatings that contain lead, cadmium, beryllium, or mercury, the use of local exhaust ventilation or an air-supplied respirator is required (general dilution ventilation is not enough). When welding or cutting stainless steel, mechanical ventilation shall be adequate to remove the fumes generated (e.g., local exhaust ventilation).

OXYGEN-FUEL GAS WELDING AND CUTTING SPECIFIC REQUIREMENTS

Oxygen-fuel gas welding or cutting involves combining a fuel gas (e.g., acetylene) with oxygen to produce a flame to melt a filler material and weld metal together or heat metal to cut through it. Typically, the oxygen/fuel gas are stored separately in cylinders. Regulators are opened on the cylinders to allow the oxygen/fuel gas to flow through hoses to a torch. The torch has valves that can be adjusted to allow the mixing of oxygen/fuel gas. When a spark is introduced to the torch with gases flowing, the gas mixture creates a flame to be used for welding or cutting. The supply of either oxygen/fuel gas can be adjusted for different operational needs. Figure 16.5 illustrates an oxygen-fuel gas welding torch.

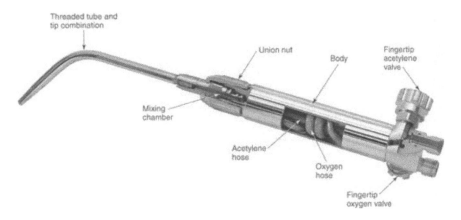

Figure 16.5

Mixtures of fuel gases and air or oxygen may be explosive and shall be guarded against to prevent fire or explosion. The most common fuel gas used for this is acetylene. Under no condition shall acetylene be generated, piped, or utilized at a pressure above 15 psig or 30 psia.

Rules, procedures, and instructions covering the operation and maintenance of oxygen/fuel gas gas supply equipment shall be readily available. All portable cylinders used for the storage and transfer of compressed gases (e.g., oxygen and acetylene) shall be legibly labeled as to their contents. Cylinders shall be kept away from sources of heat. When in storage, oxygen cylinders shall be separated from fuel gas cylinders by at least 20 feet or by a noncombustible barrier at least 5 feet high having a fire-resistance rating of at least one-half hour. Except when cylinders are in use, valve protection caps shall be in place. Cylinders must be stored and secured in a way that prevents incidental tip-over. Cylinders, valves, regulators, and hoses shall be kept free of oil, grease, or flammable materials. Cylinders must be transported in a way that prevents damage and be kept in a vertical position during transport.

ARC WELDING AND CUTTING
SPECIFIC REQUIREMENTS

It is common to use an electric arc for welding operations. An electric arc is made to pass between the metal and the filler material. The filler material is typically an electrode (stick of metal) or a wire. Figure 16.6 shows a wire welding machine. Often, electric welding can involve a shielding gas (e.g., nitrogen or argon) that helps with the quality of the weld. The electric arc creates a highly localized temperature that melts the filler material and

Figure 16.6

heats surrounding material to join metal together. Voltage and amperage are adjusted at the welder based on the materials and methods involved. Electric arc welding can be very intricate and should be respected for its difficulty, hazards, and art form.

Employees designed to operate arc welding equipment shall be qualified to do so. Electric arc welding machines must be kept in safe conditions and used per the manufacturer's instructions to prevent injuries, especially related to electric shock. Machines used for arc welding shall be designed and constructed to carry their rated electrical load and shall be suitable for operation in atmospheres containing flammable gases, dust, fumes, and light rays. Machines shall be specially designed to withstand specific conditions, if present, such as: exposure to corrosive fumes, exposure to stream/humidity, exposure to excessive oil vapor, exposure to abnormal vibration, exposure to excessive dust, and exposure to the weather.

REVIEW QUESTIONS

1. If not adequately guarded, how far away does combustible material need to be from hot work operations?
2. When is a fire watch needed for welding and what is the primary responsibility of a fire watch?
3. What is the purpose of a hot work permit? What conditions might need to be met according to a hot work permit?

4. What procedures need to be followed regarding welding on containers that contain a flammable liquid?
5. Describe the storage requirements related to the separation of oxygen and fuel-gas cylinders.

REFERENCES

Occupational Safety & Health Administration [OSHA]. (2012). Regulations (Standards-29 CFR 1910.252). Retrieved from https://www.osha.gov/laws-regs/regulations/standardnumber/1910/1910.252

Occupational Safety & Health Administration [OSHA]. (2007). Regulations (Standards-29 CFR 1910.253). Retrieved from https://www.osha.gov/laws-regs/regulations/standardnumber/1910/1910.253

Occupational Safety & Health Administration [OSHA]. (2005). Regulations (Standards-29 CFR 1910.254). Retrieved from https://www.osha.gov/laws-regs/regulations/standardnumber/1910/1910.254

https://www.nfpa.org/-/media/Files/Training/Hot-work/HotWorkPermit.ashx

APPENDIX A—HOT WORK PERMIT

HOT WORK PERMIT

Seek an alternative/safer method if possible!

Before initiating hot work, ensure precautions are in place as required by NFPA 51B and ANSI Z49.1.
Make sure an appropriate fire extinguisher is readily available.

This Hot Work Permit is required for any operation involving open flame or producing heat and/or sparks. This work includes,
but is not limited to, welding, brazing, cutting, grinding, soldering, thawing pipe, torch-applied roofing, or chemical welding.

Date	Hot work by ❑ Employee ❑ Contractor
Location/Building and floor _____	Name (print) and signature of person doing hot work
Work to be done _____ _____	I verify that the above location has been examined, the precautions marked on the checklist below have been taken, and permission is granted for this work.
Time started _____ Time completed _____	Name (print) and signature of permit-authorizing individual (PAI)
THIS PERMIT IS GOOD FOR ONE DAY ONLY	

❑ Available sprinklers, hose streams, and extinguishers are in service and operable.

❑ Hot work equipment is in good working condition in accordance with manufacturer's specifications.

❑ Special permission obtained to conduct hot work on metal vessels or piping lined with rubber or plastic.

Requirements within 35 ft (11 m) of hot work
❑ Flammable liquid, dust, lint, and oily deposits removed.
❑ Explosive atmosphere in area eliminated.
❑ Floors swept clean and trash removed.
❑ Combustible floors wet down or covered with damp sand or fire-resistive/noncombustible materials or equivalent.
❑ Personnel protected from electrical shock when floors are wet.
❑ Other combustible storage material removed or covered with listed or approved materials (welding pads, blankets, or curtains; fire-resistive tarpaulins), metal shields, or noncombustible materials.
❑ All wall and floor openings covered.
❑ Ducts and conveyors that might carry sparks to distant combustible material covered, protected, or shut down.

Requirements for hot work on walls, ceilings, or roofs
❑ Construction is noncombustible and without combustible coverings or insulation.
❑ Combustible material on other side of walls, ceilings, or roofs is moved away.

Requirements for hot work on enclosed equipment
❑ Enclosed equipment is cleaned of all combustibles.
❑ Containers are purged of flammable liquid/vapor.
❑ Pressurized vessels, piping, and equipment removed from service, isolated, and vented.

Requirements for hot work fire watch and fire monitoring
❑ Fire watch is provided during and for a minimum of 1 hour after hot work, including any break activity.
❑ Fire watch is provided with suitable extinguishers and, where practical, a charged small hose.
❑ Fire watch is trained in use of equipment and in sounding alarm.
❑ Fire watch can be required in adjoining areas, above and below.
❑ Yes ❑ No   Per the PAI/fire watch, monitoring of hot work area has been extended beyond 1 hour.

NFPA 51B

Chapter 17

Electrical Safety

INTRODUCTION AND SCOPE

Electricity can pose a variety of direct and indirect hazards to employees in industrial environments. The primary, direct hazards of electricity are electrical shock and electrocution. An electrical shock occurs when a person becomes a conductor (part of an electrical circuit). If an electrical shock causes death, it is referred to as electrocution. Electricity also presents indirect hazards. For example, if someone is shocked when working from a ladder, they can fall from the ladder and be hurt as a result. Also, faulty electrical equipment and misuse of electrical systems can cause fires. For example, when there is an increased resistance in an electrical circuit (e.g., an electrical cord compressed) over time, this can cause the buildup of heat and eventually fire. Finally, electricity can cause a major event called an arc flash (or arc blast). An arc flash occurs when electricity passes through air from one conductor to another. This can occur when there are exposed energized electrical conductors and the current jumps from the electrical system to something that was not intended to conduct the current. Arc flashes can happen when electrical equipment is compromised (e.g., corroded and covered with dust) but used as designed, or it can be a result of an external event like a tool or screw dropped on to energized conductors. Arc flashes result in a major release of pressure, heat, debris, and other hazards. Extreme care must be taken when working on or near energized electrical current.

Electrical current is the flow of electrons through a conductor. A conductor is something that allows the flow of electrons from one point to another.

Electrical current is measured in amperes (amps). Electrical current from one point to another is directly proportional to the voltage difference between the two points. This proportion includes the resistance or measure of opposition to electrical current. This relationship is known as Ohm's Law and is described in equation 17.1, where I is current, V is voltage, and R is resistance.

$$I = V / R$$

(17.1)

A common analogy for electricity is comparing the flow of electrons through a conductor (e.g., metal wire) to the flow of water through a pipe. Both the metal wire and the pipe can be considered a conductor through which energy flows. Electrical voltage is analogous to the pressure of water flowing through the pipe. Electrical resistance is analogous to the size of the pipe. A smaller pipe (or small gauge wire) has more resistance. For electricity, the type of material also affects the resistance. Electrical current is analogous to the overall flow of water through the pipe. The current that flows through a conductor depends directly on both the pressure (i.e., voltage) and resistance (i.e., size of conductor). Current, voltage, and resistance of electricity are important factors that are related closely.

When it comes to protection from electric shock and arc flash, the focus is on measures of current and voltage. For current, even very small magnitudes can be hazardous. In general, humans can sense an electrical shock at about 1 milliamp (mA). If an electrical current passes through the heart, as little as 100 mA can cause the heart to stop beating. Most electrical currents in industrial facilities are at least 15 or 20 amps (not milliamps). This is more than 10,000 milliamps. There is no question almost all electrical circuits in industrial facilities can be hazardous (even fatal) based on their current. For this reason, when assessing the risk of electrical systems, the focus is often on voltage (the greater the voltage, the more hazards present). Again, this is because virtually every industrial circuit has enough amps to be dangerous. Even very small magnitudes of current can be dangerous, but the level of risk of electrical shock depends on several factors such as the path electricity takes through the body, the amount of time a person is part of the circuit, the voltage involved, and other things. When it comes to arc flash, a hazard discussed in more detail in the following sections, the primary focus is also on the voltage of the electrical system.

In addition to understanding the parameters and hazards of electricity, it will be helpful to be familiar with the following types of electrical equipment and devices.

- Electrical panels (panel boards)—A panel or group of panels designed to be placed in a cabinet or box placed in or against a wall and accessible only from the front; for the control of light, heat, or power circuits; usually containing over-current devices (circuit breakers) that can be used to open or close circuits for control of hazardous energy (figure 17.1 shows a ~208-volt electrical panel)

- Circuit breaker—a switch (something that can open or close an electrical circuit) designed to detect an overload of electrical current that will open the electrical circuit to protect equipment from damage if an overload is detected; commonly found in electric panels (figure 17.2 shows a common 20-amp circuit breaker)
- Electrical disconnects—A device by which the conductors of a circuit can be disconnected from their source of supply; most commonly in industrial facilities, these are in the form of vertical knife switches (figure 17.3 shows a vertical knife switch disconnect)
- Receptacle—a contact device for the connection of an attachment plug (an outlet); it allows the equipment to be plugged into and receive electrical current from fixed, installed electrical circuit (figure 17.4 shows a common 15-amp receptacle)
- Junction box—a metal casing designed to house electrical conductor connections (figure 17.5 shows a common junction box with wires inside connected by wire nuts)

Figure 17.1

Figure 17.2

Figure 17.3

Figure 17.4

Electrical safety can be complicated, and there are requirements for electrical safety that come from sources other than just OSHA. State and local fire codes and building codes set standards and requirements for electrical installation, safe use of equipment, and more. Particularly, safety professionals should be familiar with the fire and building codes governing electrical safety in the state which they work in. Fire and building code requirements are not explicitly covered in this chapter; however, there is some overlap with minimum OSHA requirements and standards.

The OSHA standards covered in this chapter include:

- 29 CFR Subpart S—Electrical:
- 29 CFR 1910.303—General
- 29 CFR 1910.305—Wiring methods, components, and equipment for general use
- 29 CFR 1910.307—Hazardous (classified) locations
- 29 CFR 1910.333—Selection and use of work practices

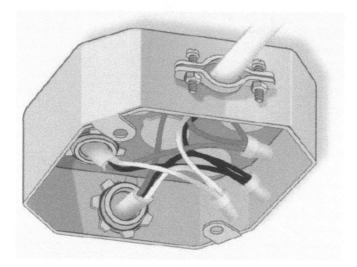

Figure 17.5

Also included in this chapter is a summary of the National Fire Protection Agency's (NFPA) Standard for Electrical Safety in the Workplace (70E)
  Not included:

* Under 29 CFR Subpart S—Electrical:
  ○ Under 29 CFR 1910.303:
    · Specific requirements for examination, installation, and use
    · Requirements for electrical connections
    · Requirements for arcing parts
  ○ 29 CFR 1910.304—Wiring design and protection
  ○ Under 29 CFR 1910.305:
    · Temporary wiring requirements applicable to electrical equipment other than flexible cords (extension cords)
    · Requirements for cable trays and open wiring on insulators
    · Requirements for covers and canopies for cabinets, boxes, and fittings
    · Requirements for switches, switchboards, and panelboards
    · Requirements for enclosures for damp or wet conditions
    · Requirements for portable cables over 600 volts
    · Requirements for motors, transformers, capacitors, and storage batteries
  ○ 29 CFR 1910.306—Specific purpose equipment and installations
  ○ 29 CFR 1910.334—Use of equipment
  ○ 29 CFR 1910.335—Safeguards for personnel protection

NARRATIVE

Jeff is an electrician and has a wife and three young children. Jeff and his family recently moved to a new home they had built. Because of money complications and hardship, Jeff could not afford to pay for the completion of the new house. For that reason, Jeff and his family were forced to move in to an unfinished home. When Jeff and his family move in, the second floor of the house is mostly bare lumber. Jeff has the skills to finish the house by himself, but he works full time and has three kids to take care. Jeff spends every minute of his free time working on the house to finish the upstairs bedrooms for his kids. Because Jeff works hard to provide for his family, and money is tight, the resources he can spend on finishing the upstairs bedrooms are very limited. Until he can finish the bedrooms, Jeff's three, small children sleep in the same queen-sized bed as him and his wife. Jeff's children are afraid to sleep anywhere else, and both Jeff and his wife would rather spend the night crowded, together as a family than sleeping on a couch. Jeff wants his wife and children to be as comfortable as possible, so he spends his sleepless nights on his side, on the far side of the bed, using just a sliver of the mattress.

It is a Friday morning, and Jeff rolls off his sliver of the bed to head in to work. Jeff is assigned to work on a 480-volt electrical system part of a motor control center. As part of the job, Jeff must access the inside of the electrical cabinet where the electrical conductors are housed. Jeff has been working as an electrician for long enough to know that opening the electrical cabinet and exposing the energized electrical conductors, even to just do a voltage test, requires compliance with safety procedures and the proper electrical protection.

Jeff prepares for equipment shutdown to work inside the electrical cabinet. While Jeff is shutting down the equipment, an operations supervisor is yelling in his ear. The supervisor is complaining that they are behind schedule, and Jeff better gets this job done quickly. Jeff is agitated, but he is trying his best. Jeff has not worked on this particular electrical circuit before. To shut power off to the electrical cabinet, he finds the circuit breaker that seems to be labeled for the cabinet he will open. Jeff opens the breaker and applies his lock and tag. In Jeff's experience at the facility, labels are generally correct, but you must verify to be sure the equipment is successfully deenergized. With the operations supervisor continuing to aggravate and rush him, Jeff skips the step of voltage testing with a multimeter to verify the conductors in the cabinet are deenergized. Jeff also fails to put on his hardhat and arc flash protective face shield. Jeff did not know the breaker was mislabeled, so Jeff did not turn off the power to the cabinet he is working in. Jeff does not know he is working in a cabinet with live 480-volt conductors without proper protection.

To complete his work, Jeff starts to remove a screw, as he does, he drops his screwdriver onto the energized conductors within the cabinet. This causes the electricity to jump to this fallen tool. A flash of an electrical arc releases a forceful blast of pressure. The force blows Jeff back against the wall behind him. His body receives major burns, and his face is badly disfigured. The impact trauma Jeff receives to his head as a result of being thrown in the arc blast is fatal. Jeff dies as a result of the blast on the cold floor of the motor control room. During Jeff's funeral, his kids cannot recognize him due to the burns on his face. The family Jeff leaves behind is forced to move to a small apartment. The children sleep on cots, and Jeff's wife spends nights crying in a cold, lonely bed. Jeff's wife would do anything to go back to the nights when their family of five slept in a crowded, queen-sized bed.

SUMMARY OF OSHA STANDARD

Electrical Equipment Acceptability and Labeling

All electrical equipment must be free from recognized hazards likely to cause death of serious harm to employees. Electrical equipment is a general term including material, fittings, devices, appliances, fixtures, apparatus, and the like, used as a part of, or in connection with, an electrical installation. All electric equipment must be acceptable for its installation. Simply speaking, acceptable for installation means tested, listed, and labeled as such by a national testing origination such as Factory Mutual (FM) or the Underwriter's Laboratory (UL). These organizations test electrical equipment for safety. Listed and labeled electrical equipment must be installed and used in accordance with any instruction in the listing and labeling. In other words, electrical equipment must be used as designed and tested. One example of a violation of this requirement is using a flexible extension cord (designed for temporary use) as permanent, fixed wiring. Figure 17.6 shows an extension cord that has been spliced and connected to fixed wiring within a junction box. The extension cord is not acceptable for this installation.

Electrical equipment (e.g., electric panels, disconnects, transformers) must be marked or labeled. The marking or label must include the manufacturer's name or other description markings. The label must also include the voltage, current, wattage, or other ratings as necessary. These markings must be durable for the environments they will be exposed to. Each disconnecting means (e.g., electrical knife switch disconnects, circuit breakers within electrical panels) for electrical motors and most electrical circuits shall be marked to indicate its purpose unless its purpose is immediately evident.

Figure 17.6

These markings shall be durable for the environments they will be exposed to. Electrical disconnects shall be capable of being locked in the open position (for lockout/tagout). Figure 17.7 depicts an electrical disconnect that is marked, labeled, and capable of accepting a lock.

REQUIREMENTS FOR ELECTRICAL EQUIPMENT OPERATING AT 600 VOLTS OR LESS

Space about Equipment

Employers shall ensure adequate clearance (space) is maintained near electrical equipment (e.g., electric panels). There shall be sufficient access and working space in front of electrical equipment and maintained around all equipment to allow ready and safe operation or maintenance of the equipment. Working space means space required for examination, adjustment, servicing, or maintenance while energized or voltage testing to verify equipment is not energized. The depth of the working space (clearance in front) required depends on the condition of the equipment and the operating voltage. OSHA Table S-1 in 29 CFR 1910 Subpart S (appendix A of this chapter) describes the prescribed clearance. However, most commonly, electrical equipment is installed such that exposed live parts are on one side and no live or grounded parts are on the other side of the working space (e.g., an electric panel on a wall). And in this case, the required working

Figure 17.7

space or clearance is 3 feet. In addition to appropriate clearance in front of electrical equipment, the width of the working space shall be the width of the electrical equipment or 30 inches, whichever is greater. In all cases, the width of clear working space must permit at least a 90-degree opening of equipment doors or hinged panels. Working space must be maintained and cleared at all times and may not be used for storage. In general, the head-room of a working space shall be no less than 6.5 feet high and at least the height of the top of the electrical equipment. Figure 17.8 illustrates the clear working space required for most electrical equipment installed on a wall. At all times, there must be at least one entrance to provide access to the clear working space.

It is worth noting that working clearance requirements (e.g., 3-feet clearance) apply to electrical disconnects that may require examination, adjustment, servicing, or maintenance while energized or voltage testing. More generally speaking, electrical disconnects must be readily accessible for use (e.g., able to be reached and used without much effort). In some cases, electrical disconnects may need to be accessed for examination, adjustment, servicing, or maintenance, and for this reason, the working space clearance requirement can apply.

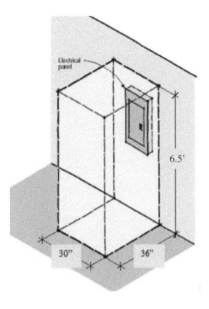

Figure 17.8

GUARDING OF LIVE PARTS

Any live (energized) parts (conductors of electricity) of electrical equipment operating at 50 volts or more shall be guarded against accidental contact by use of cabinets, guards, enclosures, or by any of the following means:

- By a room or vault only accessible by qualified persons
- By suitable permanent partitions which prevent accidental contact
- By elevation 8 feet or greater

When it comes to electrical equipment less than 600 volts, the most common method of guarding energized parts is by physical guards, covers, or enclosures.

Requirements for Electrical Equipment Operating Over 600 Volts

Generally, electrical installations of equipment over 600 volts must be in a vault, room, or closet and controlled by lock and key to be accessed by qualified persons only. If a fence is used (e.g., outdoor equipment), it shall not be less than 7-feet high. To access workspace to electrical equipment over 600 volts, there must be at least one entrance not less than 24 inches wide and

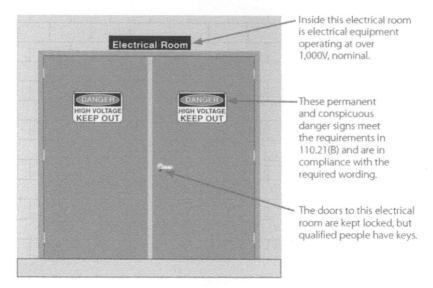

Inside this electrical room is electrical equipment operating at over 1,000V, nominal.

These permanent and conspicuous danger signs meet the requirements in 110.21(B) and are in compliance with the required wording.

The doors to this electrical room are kept locked, but qualified people have keys.

Figure 17.9

6.5-feet high. The minimum working space depth (clearance in front) shall not be less than the distances specified in Table S-2 of 29 CFR 1910 Subpart S. This table is provided as appendix B of this chapter. Just like for electrical equipment less than 600 volts, the clearance to equipment greater than 600 volts depends on the condition of the electrical equipment and the operating voltage. Entrances to enclosures or rooms containing electrical equipment over 600 volts shall be kept locked at all times with access only by qualified persons. The entrance to the enclosure or room shall have a warning sign reading: "DANGER—HIGH VOLTAGE—KEEP OUT." Figure 17.9 shows an electrical room with required signage on the door.

WIRING DESIGN, METHODS AND PROTECTION, AND GENERAL EQUIPMENT REQUIREMENTS

OSHA sets many requirements related to "wiring design and protection" and "wiring methods, components, and equipment." These standards can be fairly complicated. OSHA recognizes that electrical installation must comply with the National Electrical Code (NEC) which also sets design, methods, and installation requirements for electrical equipment. These specific requirements are quite technical and specific. In general, young safety professionals do not have to know all the technical requirements (or be knowledgeable as an electrician), but they should be able to identify common OSHA violations and hazards related to electrical equipment. This section intends to present only the most common, need-to-know elements of the OSHA standards concerning wiring design, methods, and protection as well as requirements for general equipment (e.g., appliances and motors).

TEMPORARY WIRING AND FLEXIBLE CORDS

Flexible cords used as temporary wiring (i.e., extension cords) is used for short-term power supply. It is not fixed wiring. It is not installed for permanent use (e.g., inside conduit). The primary example of temporary wiring is an extension cord, referred to as a flexible cord used as temporary wiring inside the OSHA standard. Temporary wiring (e.g., extension cords) can only be used as follows:

- During remodeling, maintenance, repair, and similar activities
- For a period not to exceed 90 days for decorative lighting, carnivals, or similar occasions
- For experimental or developmental work, or emergencies

Temporary wiring shall be removed immediately upon completion of the project or the purpose it was installed. So, you cannot use an extension cord for a long-term power supply (it cannot be plugged in all the time, continuously, without removal). Extension cords must be used for short-term applications (e.g., to use a power tool for a short time) and then removed after use. Temporary electrical installations of more than 600 volts are only allowed for testing, experiments, emergencies, or construction. Flexible cords and cables shall be protected from accidental damage, for example, from sharp corners and pinch points.

Flexible cords and cables (e.g., extension cords) shall not be used:

- As a substitute for the fixed wiring of a structure
- Where run through holes in walls, ceilings or floors
- Where run through doorways, windows, or similar openings
- Where attached to building surfaces
- Where concealed behind building walls, ceilings, or floors
- In cable raceways (racks to hold wiring)

Flexible cords may only be used in continuous lengths without splice (they cannot be severed with wires reconnected). Flexible cords shall be arranged

Figure 17.10

Figure 17.11

so that there is no strain on the cord. Figures 17.10, 17.11, and 17.12 show various violations related to extension cord use.

Extension cords shall not be used if damaged. The two most common violations are a missing grounding prong (figure 17.13) and damaged insulation (figure 17.14). An extension cord shall be removed from service if the grounding prong is missing or insulation is damaged. In most cases, an extension cord shall not be used if the electrical tape is used to repair damaged

Figure 17.12

insulation. This is because, in most cases, using electrical tape does not return the outer insulation (outer sheath) to its original manufactured properties and characteristics.

There are two common applications of flexible cords that are designed for long term use—a surge protected power strip and a relocatable power tap (similar to a power strip, designed to be extension of building circuit) These flexible cord devices can be left plugged in if used as designed, used to power portable appliances, and rules for use of flexible cords are followed. Always refer to the manufacturer's instructions and limitations.

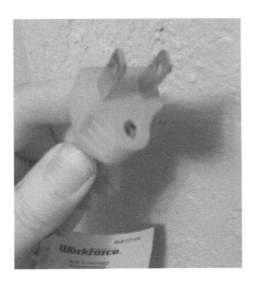

Figure 17.13

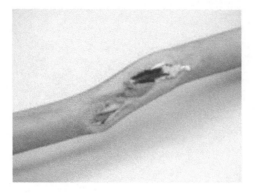

Figure 17.14

ELECTRICAL CABINETS, BOXES, AND FITTINGS

Conductors entering electrical cabinets, boxes, and fittings shall be protected from abrasion (being cut by the edges of the enclosure). This protection is commonly referred to as bushing. Figure 17.15 shows a missing bushing that causes the insulated conductor to be at risk of abrasion from the edges of the metal. It is also possible the wiring shown in figure 17.15 are flexible cables not designed or acceptable to be used as fixed wiring. Any unused openings in electrical cabinets and boxes shall be effectively closed. A common violation related to openings in electrical equipment is a missing knockout on junction boxes and other electrical enclosures. Figure 17.16 shows a missing knockout

Figure 17.15

Figure 17.16

that would be an OSHA violation. Another common violation related to openings in electrical panels is missing circuit breakers inside an electrical panel. If a circuit breaker is removed from an electrical panel, there cannot be an opening that exposes live conductors. This opening shall be filled with a covering referred to as a blank. Figure 17.17 depicts a missing circuit breaker in an electrical panel without a blank in place.

Figure 17.17

CONDUCTORS FOR GENERAL WIRING

Generally, all conductors (wires) for general wiring must be insulated. The conductor insulation must be the appropriate material based on the voltage, operating temperature, and location of use. The insulated conductors must be distinguishable by appropriate color to be able to tell which conductors are grounded and which are ungrounded. In the United States, a simple single-phase (one hot wire) 240 volt or less circuit is color-coded as follows: the ground wire is insulated in green, the hot (live) wire is insulated in black, and the neutral wire is insulated in white. Wiring color-coding can be much more complicated with multiple phases and hot wires.

EQUIPMENT FOR GENERAL USE

Light fixtures, lamp holders, and receptacles (outlets) shall not have any live parts normally exposed for accidental contact. Any electrical appliance shall not have unguarded live parts. All attachment plugs shall not have any exposed current-carrying parts except for the blades and grounding prong. Cords of plugs shall not be used with damaged insulation. And similar to flexible extension cords, for flexible cords of appliances and fixtures, electrical tape is not typically an acceptable repair. Any receptacle used in wet or damp location must

Figure 17.18

Figure 17.19

be designed for such use. If installed outdoors, a receptacle must have a weatherproof enclosure unless in a location protected by weather (e.g., under porch).

In some situations, a ground-fault circuit interrupter (GFCI) must be used. This is a device that senses if less current travels back to the electrical source than what is put out. In other words, a GFCI can detect if something (a person) becomes a conductor (an electrical shock occurs). When this decrease in current is detected, the GFCI opens the circuit to prevent the further supply of electrical energy (stops the electrical shock). Building code will require GFCI protection in damp locations and near water sources in industrial and residential buildings (e.g., in bathrooms, near sinks). Refer to your local code for specific requirements. Figure 17.18 shows a GFCI receptacle. A good rule of thumb is GFCI-equipped receptacles should be installed if within 5 feet of a water source like a water fountain or sink. GFCI protection can also be achieved at the electrical panel which feeds the receptacle. A circuit breaker

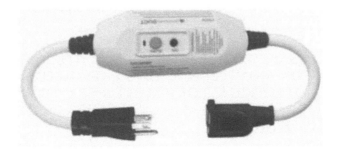

Figure 17.20

with GFCI protection is shown in figure 17.19. Whether equipped at the receptacle or breaker, a GFCI device will have a test button to test the switch. All 120-volt 15- and 20-amp receptacle outlets on construction sites, which are not a part of the permanent wiring of the building or structure and which are in use by employees, shall have GFCI protection. This is typically achieved using a cord connector. A GFCI-equipped cord connector is depicted in figure 17.20.

HAZARDOUS (CLASSIFIED) LOCATIONS

Electrical equipment and wiring found in classified hazardous locations must meet specific requirements. Hazardous locations are those that contain flammable liquids/vapors/gases, combustible dust, or ignitable fibers. Hazardous locations are classified and defined below.

- Class I—Flammable liquids, vapors, or gases
  - Division 1—present during normal, usual conditions
  - Division 2—present only in abnormal conditions (i.e., normally contained)
- Class II—Combustible dust
  - Division 1—present during normal, usual conditions
  - Division 2—present only in abnormal conditions
- Class III—Ignitable fibers
  - Division 1—present during normal, usual conditions
  - Division 2—present only in abnormal conditions

Equipment, wiring methods, and installations of equipment in classified, hazardous locations shall be listed and approved for the hazardous location. This means they must be designed as such to not be an ignition source. Equipment shall be marked to show the class for which it is approved. OSHA provides further requirements for electrical equipment installed in these locations, but the main idea to know is wiring must be designed and approved for the class and division it is used in, if it is used in a classified hazardous location.

SAFE WORK PRACTICES FOR WORK ON AND APPROACHING ENERGIZED ELECTRICAL EQUIPMENT (NFPA 70E)

NFPA 70E Overview

When it comes to working on or working near energized electrical equipment, OSHA points to the NFPA Standard for Electrical Safety in the Workplace

(NFPA 70E). This is the standard that outlines precautions for working on and near energized electrical equipment. The purpose of NFPA 70E is to control the hazards related to electrical shock and arc flash. An electric arc is when current flows through air between two conductors. Arc flashes can be deadly. They result in an explosive release of pressure, high temperatures, molten metal, flying debris, UV light, and noise. Electrical equipment and its conductors are considered live until the circuit is deenergized, isolated, and verified to be deenergized. NFPA 70E is based on the protective strategies of establishing an electrically safe working environment, planning tasks with an Electrical Safety Program, selecting appropriate personal protective equipment (PPE), and training workers on safe work practices. The rest of this chapter summarizes provisions of NFPA 70E.

NFPA70E APPLICATION

OSHA considers exposed electrical less than 50 volts to be non-hazardous. Exposures to circuits less than 50 volts do not require deenergization, and therefore NFPA70E does not apply. Electrical work on or approaching energized systems 50 volts or greater requires certain safeguards and work practices (i.e., compliance with reference standard NFPA 70E). Electrical work usually falls into one of the following categories, and depending on the category, may require a permit to authorize work:

1. Electrically safe working condition (deenergized, locked/tagged out, and verified to be deenergized)
   - Does not require a permit to authorize work
2. Voltage testing or diagnostic work with appropriate PPE
   - Does not require a permit to authorize work
3. Work on energized electrical
   - Requires a permit to authorize work

It is important to note the electrical circuit or equipment is considered to be energized until the circuit is locked and tagged out (LOTO) and the absence of energy is verified. Deenergization and lockout tagout to achieve an electrically safe work environment must always be the first, preferred approach to electrical work. After deenergization and LOTO are applied, it is common to conduct voltage testing to verify deenergization was successful. During voltage testing, because deenergization has not yet been verified, NFPA 70E provisions (e.g., required PPE) apply. For example, if an electrician shuts off power to equipment and then must use a multimeter to do voltage testing inside the equipment to verify the absence of energy, this voltage testing task is considered live electrical work until verification is made that there is no

Figure 17.21

hazardous energy present. A multimeter is a device that measures electrical properties such as current and voltage. Typically, the use of a multimeter involves touching two probes to the electrical conductors. If voltage is zero (across all hot conductors [phase to phase] and all phases to ground), the equipment is deenergized. In this section, electrical work means work on live, energized conductors, and voltage testing to verify the absence of energy. Figure 17.21 illustrates voltage testing to verify the absence of energy that would be considered live electrical work.

NFPA 70E ELECTRICAL WORK ASSESSMENT

Electrical work hazards and the need for arc flash PPE are assessed using NFPA70E standard tables and charts. The standard first requires identification of the electrical task. Examples of tasks include opening hinged doors and panels of electrical equipment to reveal unguarded energized conductors, voltage testing, and other electrical tasks. Essentially, electrical work involves any exposure to energized electrical of 50 volts or greater. This includes simply removing panels, covers, or guards that reveal energized conductors, voltage testing, and actual work on electrical equipment. Operating electrical equipment switches in good condition, such as electrical disconnects and circuit breakers, do not require arc flash PPE unless the safety of the equipment has been compromised (e.g., equipment not in good condition).

Based on the task being performed, the table and charts will direct if arc flash PPE is required. Then, the voltage of the circuit to be worked on must be used to identify a limited approach boundary and a restricted approach boundary. A limited approach boundary is the distance at which unqualified electrical persons may not encroach. A restricted approach boundary is the distance at which only qualified persons are allowed, where insulated tools and shock protection (voltage rated gloves) are required, and where work inside required a permit (unless voltage testing/diagnostic work). Next, using the voltage of the circuit to be worked on and the type of equipment to be worked on, an arc flash boundary is identified. An arc flash boundary is the distance inside which arc-rated PPE must be worn. The level of arc flash PPE is then identified based on the type of equipment to be worked on. Arc flash PPE is divided into four categories. The higher the category of PPE required,

Hazard/Risk Category 4 cal/cm²	1	Arc-rated long-sleeve shirt Arc-rated pants or overall Arc-rated face shield with hard hat Safety glasses Hearing protection Leather & voltage rated gloves (as needed) Leather work shoes
Hazard/Risk Category 8 cal/cm²	2	Arc-rated long-sleeve shirt Arc-rated pants or overall Arc-rated face shield & balaclava or Arc flash suit with hard hat Safety glasses, Hearing protection Leather & voltage rated gloves (as needed) Leather work shoes
Hazard/Risk Category 25 cal/cm²	3	Arc-rated long-sleeve jacket Arc-rated pants Arc-rated flash hood with hard hat Safety glasses, Hearing protection Leather & voltage rated gloves (as needed) Leather work shoes
Hazard/Risk Category 40 cal/cm²	4	Arc-rated long-sleeve jacket Arc-rated pants Arc-rated flash hood with hard hat Safety glasses, Hearing protection Leather & voltage rated gloves (as needed) Leather work shoes

Figure 17.22

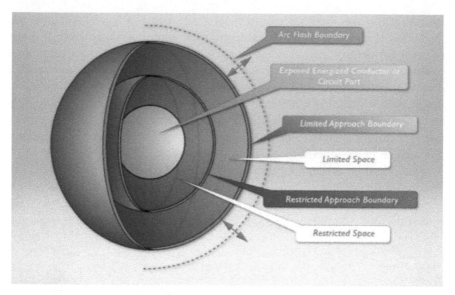

Figure 17.23

the higher the arc flash resistance must be achieved with the clothing and protective equipment. Arc flash resistance is expressed in cal/cm^2 (energy per area). Figure 17.22 shows the different categories of arc flash PPE. Figure 17.23 illustrates the NFPA 70E approach boundaries.

NFPA 70E ELECTRICAL WORK PERMIT

Work on energized electrical 50 volts or greater, inside the restricted approach boundary (other than voltage testing and diagnostic work) requires written authorization via a work permit. Work on energized electrical (other than testing/diagnostics) is only allowed if deenergization would create a hazard or if deenergization is infeasible, for example, if shutting down equipment would shut off life support system in a hospital. There are not many examples of when deenergization is infeasible or when deenergization would create more of a hazard. Deenergization and LOTO are always the preferred approach, so a permit is not needed. For this reason, energized electrical work permits should be uncommon in general industry. An electrical work permit has standard sections. First, the permit will include general information about the task such as a description of the equipment to be worked on, a description of the tasks, justification of why the task must be performed live, and documentation of a job briefing with employees involved. Then, the permit should include a hazard assessment that involves identifying approach boundaries and what PPE is required. Finally, the permit should include a list of equipment that must be used to maintain safety and a section for written approval from a qualified supervisor.

NFPA 70E TRAINING

Personnel performing electrical work (including voltage testing to verify circuits are deenergized) must be qualified. An employee may be qualified for some tasks/equipment but not others. A qualified person:

- Has the skills and techniques to distinguish exposed energized parts from other electrical equipment
- Has the skills and technique to determine the nominal voltage
- Is knowledgeable about approach distances
- Is knowledge about the decision-making process necessary to determine the degree and extent of the hazard
- Knows what PPE is needed based on the electrical task

All qualified personnel who work on electrical system must be trained on NFPA70E requirements and the provisions of the employer's Electrical Safety Program. Qualified persons must also be trained on the control of hazardous energy (lockout/tagout), emergency procedures related to electrical incidents, and first aid/CPR.

ELECTRICAL PROTECTIVE EQUIPMENT

OSHA sets requirements for types of electrical insulated PPE such as rubber gloves and sleeves. Tasks that require such electrical PPE depend on an adequate hazard assessment of exposures, work tasks, and the environment. Generally speaking, any task that falls under NFPA70E (work inside the restricted approach boundary) requires shock protection and at minimum rubber insulated gloves. This section will focus on insulated gloves for electrical shock protection. Insulated gloves shall be classified and marked as either Class 00, 0, 1, 2, 3, or 4. Appendix C of this chapter provides the OSHA table that identifies the maximum use voltage for each class of electrical insulated equipment. For example, if working on an electrical circuit less than 500 volts, Class 00 gloves would be appropriate.

Insulated gloves must be inspected before each use to ensure they are free from damage, corrosion, or other conditions that may compromise their effectiveness. Insulated gloves must be given an air test before use. This is done by the employee blowing air into the glove, sealing it, and ensuring pressure is not loss that would indicate the presence of a hole or leak. Rubber insulated gloves must be electrically tested for the ability to insulate from electrical shock every six months. This is typically done by sending gloves to a company to complete the testing.

REVIEW QUESTIONS

1. Define electrical current, voltage, and resistance. What are the most important factors when assessing the risk of shock from an electrical circuit? How many amps can stop the heart?
2. Name two indirect hazards of electricity.
3. What is the most commonly working space required in front of electrical equipment operating at less than 600 volts? What is working space?
4. Describe five OSHA violations related to extension cord use.
5. At what voltage threshold must expose energized conductors be protected from incidental contact?
6. Describe how a GFCI works and its purpose.
7. What is the purpose of NFPA 70E?
8. Regarding NFPA 70E: What information is needed to conduct a hazard assessment and identify required PPE? Describe all three approach boundaries. When is a permit required, and when is it not? What is the first, preferred approach to electrical work? Does voltage testing on conductors greater than 50 volts require arc flash PPE? What type of training is required for a qualified person who performs electrical work and voltage testing under NFPA 70E?

REFERENCES

Occupational Safety & Health Administration [OSHA]. (2008). Regulations (Standards-29 CFR 1910.303). Retrieved from https://www.osha.gov/laws-regs/regulations/standardnumber/1910/1910.303

Occupational Safety & Health Administration [OSHA]. (2007). Regulations (Standards-29 CFR 1910.305). Retrieved from https://www.osha.gov/laws-regs/regulations/standardnumber/1910/1910.305

Occupational Safety & Health Administration [OSHA]. (2007). Regulations (Standards-29 CFR 1910.307). Retrieved from https://www.osha.gov/laws-regs/regulations/standardnumber/1910/1910.307

Occupational Safety & Health Administration [OSHA]. (1994). Regulations (Standards-29 CFR 1910.333). Retrieved from https://www.osha.gov/laws-regs/regulations/standardnumber/1910/1910.333

Occupational Safety & Health Administration [OSHA]. (1990). Regulations (Standards-29 CFR 1910.334). Retrieved from https://www.osha.gov/laws-regs/regulations/standardnumber/1910/1910.334

Occupational Safety & Health Administration [OSHA]. (1990). Regulations (Standards-29 CFR 1910.335). Retrieved from https://www.osha.gov/laws-regs/regulations/standardnumber/1910/1910.335

(NFPA) 70E—Standard for Electrical Safety in the Workplace

APPENDIX A—OSHA 29 CFR 1910 SUBPART S TABLE S-1 CLEAR WORKING SPACE FOR ELECTRICAL EQUIPMENT 600 VOLTS OR LESS

Table S-1—Minimum Depth of Clear Working Space at Electric Equipment, 600 V or Less

Nominal voltage to ground	Minimum clear distance for condition[23]					
	Condition A		Condition B		Condition C	
	m	ft	m	ft	m	ft
0-150	[1]0.9	[1]3.0	[1]0.9	[1]3.0	0.9	3.0
151-600	[1]0.9	[1]3.0	1.0	3.5	1.2	4.0

Notes to Table S-1:

1. Minimum clear distances may be 0.7 m (2.5 ft) for installations built before April 16, 1981.

2. Conditions A, B, and C are as follows:

Condition A—Exposed live parts on one side and no live or grounded parts on the other side of the working space, or exposed live parts on both sides effectively guarded by suitable wood or other insulating material. Insulated wire or insulated busbars operating at not over 300 volts are not considered live parts.

Condition B—Exposed live parts on one side and grounded parts on the other side.

Condition C—Exposed live parts on both sides of the work space (not guarded as provided in Condition A) with the operator between.

3. Working space is not required in back of assemblies such as dead-front switchboards or motor control centers where there are no renewable or adjustable parts (such as fuses or switches) on the back and where all connections are accessible from locations other than the back. Where rear access is required to work on deenergized parts on the back of enclosed equipment, a minimum working space of 762 mm (30 in.) horizontally shall be provided.

APPENDIX B—OSHA 29 CFR 1910 SUBPART S TABLE S-2 CLEAR WORKING SPACE FOR ELECTRICAL EQUIPMENT OVER 600 VOLTS

Table S-2—Minimum Depth of Clear Working Space at Electric Equipment, Over 600 V

Nominal voltage to ground	Minimum clear distance for condition[23]					
	Condition A		Condition B		Condition C	
	m	ft	m	ft	m	ft
601-2500 V	0.9	3.0	1.2	4.0	1.5	5.0
2501-9000 V	1.2	4.0	1.5	5.0	1.8	6.0
9001 V-25 kV	1.5	5.0	1.8	6.0	2.8	9.0
Over 25-75 kV[1]	1.8	6.0	2.5	8.0	3.0	10.0
Above 75 kV[1]	2.5	8.0	3.0	10.0	3.7	12.0

Notes to Table S-2:

[1]Minimum depth of clear working space in front of electric equipment with a nominal voltage to ground above 25,000 volts may be the same as that for 25,000 volts under Conditions A, B, and C for installations built before April 16, 1981.

[2]Conditions A, B, and C are as follows:

Condition A—Exposed live parts on one side and no live or grounded parts on the other side of the working space, or exposed live parts on both sides effectively guarded by suitable wood or other insulating material. Insulated wire or insulated busbars operating at not over 300 volts are not considered live parts.

Condition B—Exposed live parts on one side and grounded parts on the other side. Concrete, brick, and tile walls are considered as grounded surfaces.

Condition C—Exposed live parts on both sides of the work space (not guarded as provided in Condition A) with the operator between.

[3]Working space is not required in back of equipment such as dead-front switchboards or control assemblies that has no renewable or adjustable parts (such as fuses or switches) on the back and where all connections are accessible from locations other than the back. Where rear access is required to work on the deenergized parts on the back of enclosed equipment, a minimum working space 762 mm (30 in.) horizontally shall be provided.

APPENDIX C—OSHA TABLE I-4 RUBBER INSULATING EQUIPMENT VOLTAGE REQUIREMENTS

Class of equipment	Maximum use voltage[1] AC rms	Retest voltage[2] AC rms	Retest voltage[3] DC avg
00	500	2,500	10,000
0	1,000	5,000	20,000
1	7,500	10,000	40,000
2	17,000	20,000	50,000
3	26,500	30,000	60,000
4	36,000	40,000	70,000

Chapter 18

Hazard Communication

Chapter 18 was written by Paige Laratonda.

INTRODUCTION AND SCOPE

Hazard Communication (HazCom) refers to the communication of chemical hazards that may be present in the workplace. The information about chemicals and their hazards must be passed down from the manufacturer, to the distributors, to employers, and to employees. Chemicals are found in almost every work setting. For example, paint may be used for an office renovation, solvents and flammable liquids can be used in an industrial plant, and corrosive acids are commonly used in chemical laboratories. Different chemicals pose different risks based on their composition, hazards, and other characteristics. In most cases, it is not a concern to get water-based paint on your hands, but for other chemicals, a relatively small amount on your skin can destroy tissue. By simply looking at a chemical or substance, one might not be able to tell if it is essentially harmless or very dangerous. Because of this, OSHA requires workers have the "right to know" about chemicals and substances in their workplace and how to protect themselves from the hazards they pose.

OSHA's HazCom requirements are intended to be consistent with the United Nations Globally Harmonized System (GHS). This means that since 2015, all employers across the globe are expected to meet the same requirements regarding the communication of chemical hazards. Before the GHS was mandated by the United Nations, each country could have its own set of rules. The GHS ensures that the chemicals distributed or supplied from any place on the map have hazard information that is clearly presented and understood to those in all countries.

The GHS brought global consistency to the classification of chemical hazards and the dissemination of chemical safety information, and the important effect of this is employer requirements under OSHA's HazCom standard. The trickle-down of chemical information to all affected employees in a workplace is to be accomplished through a comprehensive HazCom Program. A HazCom Program must include a chemical inventory, procedures for container labeling and other forms of warnings, procedures for maintaining safety data sheets (SDSs), and employee training. A chemical inventory is a list of all the hazardous chemicals employees might be exposed to. A container label provides immediate information about the chemical inside. An SDS is a document that provides technical safety information about the chemical it describes. The procedures within a written HazCom Program are essential to eliminate uncertainty about chemical risk in the work settings and crucial for preventing incidents related to chemical exposure.

Parallel to OSHA's HazCom standard, laboratories that mix and manipulate chemicals must maintain a Chemical Hygiene Plan. Laboratories covered by the OSHA laboratory safety standard are exempt from most of the requirements of the HazCom standard. A Chemical Hygiene Plan is similar to a HazCom Plan, but it's more applicable to operations of a laboratory setting. The OSHA laboratory standard is not covered in this chapter.

The OSHA standards covered in this chapter include:

- 29 CFR 1910.1200—Hazard Communication

Not included:

- Under 29 CFR 1910.1200—Hazard Communication
  - Procedures for classifying the potential physical and health hazards of chemicals during chemical production (applies to chemical manufacturers)
  - Requirements that apply specifically to distributors of chemicals (applies to those delivering, supplying, or transporting chemicals)
  - Requirements for facilities where chemicals are only handled in sealed containers that are not opened under normal conditions of use (applies to warehousing, retail sales, etc.)

NARRATIVE

Doug and Jane are two hard-working, intelligent chemical technicians who work night and day in a chemical processing department of a mineral extracting plant. Doug and Jane have worked together for a long time and are great friends. The plant requires the chemical processing department to maintain a Hazard Communication Plan, ensure proper labeling, and file safety data sheets as necessary. Both Jane and Doug receive HazCom training and are aware of the HazCom OSHA requirements.

The two competent technicians conduct standard procedures with chemicals to complete their work. A standard technique involves dissolving sedimentary rock with hydrofluoric acid (HF). HF is a highly corrosive acid. Doug is conducting this task one late evening as his stomach is growling, and his eyes grow heavy from a long week. Doug typically conducts this task to save Jane from having to handle HF, since it is a highly hazardous chemical. Jane is not familiar with HF and its hazards. Doug's wife texts him to check-in and tells him leftover spaghetti would be waiting for him when he got home. Lately, Doug has been feeling bad about all of the extra time he has been spending working optional overtime, so he decides to rush home to surprise his wife and have supper with her.

As Doug is rushing to finish up, he yells over to Jane that he would be leaving soon and asks her to please clean up after him, just this once. She agrees knowing he would do the same. Doug leaves his workstation along with an unlabeled, secondary container of HF setting on the lab bench. Doug assumed, at the beginning of the day, he would use up all of the HF in the container before the end of his shift. And since he would be the only one to use it, he did not label the container.

Doug and Jane say their goodbyes, and Doug leaves the facility. As promised, Jane immediately walked over to Doug's station to clean up and noticed an unlabeled container. Jane was shocked, because she knows Doug is typically very careful about labeling and safety. And for this reason, she mostly assumes the chemical must be harmless. But to be sure, and to verify how to dispose of it properly, Jane quickly grabs for her phone to call Doug. As she turns around, she bumps the unlabeled container of hydrofluoric acid causing it to spill onto Jane's leg. Since she does not work with the HF, she is not as aware of the hazards and is not wearing the appropriate protective equipment. She can feel a burning sensation on her leg rather quickly and instantly knows to get to the emergency shower. She cancels her call with Doug and calls 911 while water is drenching her leg. Jane is unable to identify what the chemical is that spilled on her leg but knows the pain is growing worse. Jane is taken to the hospital for severe and worsening chemical burns, but it takes a while to get there, and the hospital needs to identify the unknown chemical exposure. Luckily Jane survives, but her left leg is amputated near the hip due to extreme degradation of skin, muscle, and even bone.

Doug is at home, when he is informed over the phone about Jane's hospitalization and questioned about the use of HF in the lab. A deep feeling of regret overcomes Doug thinking about his failure to label the container of

HF. For as long as he lives, Doug thinks about costing his friend her leg and wishes he labeled the dangerous chemical. If he did, Jane could have been more prepared with appropriate PPE and received more swift medical treatment. Jane had a right to know.

SUMMARY OF OSHA STANDARD

Chemical Hazards and Hazard Communication Program Overview

A phrase often associated with HazCom is the "right to know." This phrase sums up the motive behind the HazCom standard in the workplace. It means that all employees have the right to know what physical and health hazards they may potentially be exposed to while doing their job. Examples of specific hazardous chemical use in the workplace are endless. A chemical is hazardous if it poses any potential physical or health hazards. Criteria for categorizing hazards can be found in 29 CFR 1910.1200 appendices A and B. Specific criteria are omitted from this chapter.

Examples of chemical health hazards include but are not limited to:

• Acute (immediate) toxicity
• Skin corrosion or irritation
• Eye damage or irritation
• Respiratory or skin irritation or sensitization (allergic/internal reaction)
• Mutagenicity (causes changes in DNA)
• Carcinogenicity (causes cancer)
• Reproductive toxicity
• Target organ toxicity from short-term exposure
• Target organ toxicity from long-term exposure

Examples of physical-chemical hazards include:

• Explosives
• Flammable solids, liquids, gases
• Oxidizers (increase fire risk)
• Pressure
• Reactivity

All industrial worksites that utilize hazardous chemicals are required to comply with this standard and maintain a written HazCom Program (with few exceptions). The employer shall identify and designate a qualified person to manage the program and its requirements. This employee is responsible for developing a chemical inventory, organizing SDSs, setting up HazCom training, and so on. Primary provisions under the HazCom Program include procedures for container labeling, maintaining SDSs, compiling a chemical inventory, and providing employees training.

To summarize, employers are required to develop, implement, and maintain a written HazCom Program that meets the following:

- Includes a list of the hazardous chemicals known to be present in the workplace.
- The methods the employer will use to inform employees of the hazards (e.g., labels and SDSs).
- Proof that the employer has trained each exposed individual on the potential chemical hazards present that they will be exposed to
- Is available, upon request, to employees at any given time.

Where employees must travel between workplaces during a work shift (e.g., their work is carried out at more than one geographical location), the written HazCom Program may be kept at the primary workplace facility.

Hazardous Chemical Inventory

Employers are required to assess their workplace for hazardous chemicals. All employers who use or store hazardous chemicals must compile an inventory of the substances. The inventory, at minimum, must include the name of the chemical (as described on the SDS and label) and the manufacturer of the chemical. A chemical inventory must be immediately available and is typically kept with SDSs.

Labels

Labels are important devices to provide immediate information to employees about what is inside a container and what its primary hazards are. Labels must be present and legible on all chemical containers. Figure 18.1 shows an OSHA—required label that must be present on chemical containers that come from the manufacturer (primary containers). Employers and employees shall ensure labels do not get removed or defaced. The labels of primary containers must include the following components:

- A product identifier—the name used for a hazardous chemical as it appears on the SDS. It provides a unique means by which the user can identify the chemical. The product identifier used shall permit cross-references

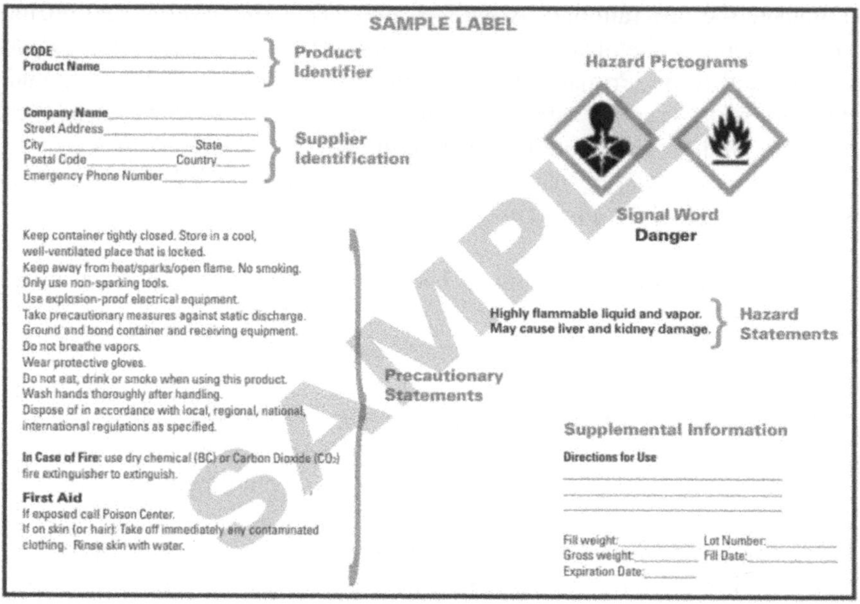

SAMPLE LABEL

CODE _____
Product Name _____   } Product
                                    Identifier

Company Name _____
Street Address _____
City _____ State ____
Postal Code _____ Country ____   } Supplier
Emergency Phone Number _____   Identification

Keep container tightly closed. Store in a cool,
well-ventilated place that is locked.
Keep away from heat/sparks/open flame. No smoking.
Only use non-sparking tools.
Use explosion-proof electrical equipment.
Take precautionary measures against static discharge.
Ground and bond container and receiving equipment.
Do not breathe vapors.
Wear protective gloves.
Do not eat, drink or smoke when using this product.
Wash hands thoroughly after handling.
Dispose of in accordance with local, regional, national,
international regulations as specified.

In Case of Fire: use dry chemical (BC) or Carbon Dioxide (CO₂)
fire extinguisher to extinguish.

First Aid
If exposed call Poison Center.
If on skin (or hair): Take off immediately any contaminated
clothing. Rinse skin with water.

Precautionary
Statements

Hazard Pictograms

Signal Word
Danger

Highly flammable liquid and vapor.   } Hazard
May cause liver and kidney damage.   Statements

Supplemental Information

Directions for Use
_____
_____
_____

Fill weight: _____   Lot Number: _____
Gross weight: _____   Fill Date: _____
Expiration Date: _____

Figure 18.1

to be made among the list of hazardous chemicals required in the written HazCom Program, the label, and the SDS. A numbering system can be used if employees are properly trained, but that is not common.

- A signal word—a word used to indicate the relative level of severity of hazard and alert the reader to a potential hazard on the label. The signal words used are "danger" and "warning." "Danger" is used for the more severe hazards, while "warning" is used for the less severe.
- Hazard statement—a statement assigned to a hazard class and category that describes the nature of the hazard(s) of a chemical, including, where appropriate, the degree of hazard. A hazard class means the nature of physical or health hazards (e.g., flammable solid, carcinogen, and oral acute toxicity). A hazard category is a division within the class. For example, Category 1 Target Organ Toxicity is most dangerous and Category 3 Target Organ Toxicity is relatively less dangerous.
- Pictogram—an illustration that may include a symbol plus other graphic elements, such as a border, background pattern, or color, that is intended to convey specific information about the hazards of a chemical. Nine pictograms are designated under this standard for application to a hazard category. Figure 18.2 illustrates the nine pictograms.
- Precautionary statement—a phrase that describes recommended measures that should be taken to minimize or prevent adverse effects

HCS Pictograms and Hazards

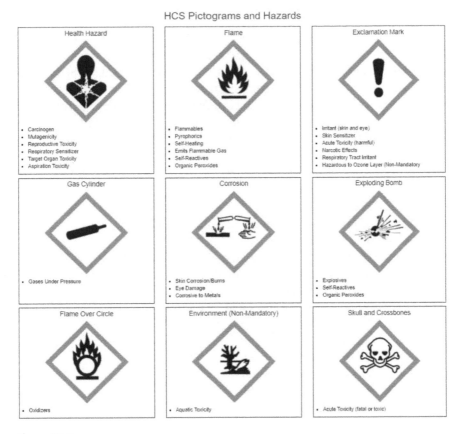

Figure 18.2

resulting from exposure to a hazardous chemical, or improper storage or handling.

• Name, address, and phone number of the chemical manufacturer, importer, or other responsible parties.

Labels must be found on both primary and secondary containers. If you pour a chemical from its original container into a different container for storage, the different container is called a secondary container. A secondary container must be labeled with at least the product identifier and hazard information (e.g., written with a sharpie). There is only one exception to labeling. Secondary containers that are used for the IMMEDIATE transfer of chemicals that are being directly HANDLED by a single employee do not need a label. This transfer container must be immediately emptied of its contents during its use for the exception to apply.

Safety Data Sheets (SDSs)

Employers are required to provide access to SDSs for each hazardous chemical that they use or store. Employers must ensure SDSs are readily accessible (able to obtain quickly and without much effort) for each chemical an employee may be exposed to during each work shift. Electronic access and other alternatives to maintaining paper copies of the SDSs are permitted as long as no barriers to immediate employee access in each workplace are created by such options (e.g., power outage).

Each SDS must contain the following:

- Section 1, Identification
- Section 2, Hazard identification
- Section 3, Composition/Information of ingredients
- Section 4, First-aid measures
- Section 5, Fire fighting measures
- Section 6, Accidental release measures
- Section 7, Handling and storage
- Section 8, Exposure controls/personal protection
- Section 9, Physical and chemical properties
- Section 10, Stability and reactivity
- Section 11, Toxicological information
- Section 12, Ecological information
- Section 13, Disposal considerations
- Section 14, Transport information
- Section 15, Regulatory information
- Section 16, Other information, including date of preparation or last revision

If the SDS is not provided with a shipment of a hazardous chemical to the employer, the employer shall obtain one from the chemical manufacturer or importer as soon as possible, and the chemical manufacturer or importer shall also provide employers with an SDS upon request. Most SDSs are found readily available online with an easy internet search; however, that is not always the case. It is important that the SDS found online is correct and from the same manufacturer. Where employees must travel between workplaces during a work shift (e.g., their work is carried out at more than one geographical location), the SDSs may be kept at the primary workplace facility. In this situation, the employer shall ensure employees can immediately obtain the required information in an emergency. SDSs are crucial documents for reference to chemical information, hazards, and precautionary measures.

Exemption for Household Products

Certain, specific substances are exempt from the HazCom Program, because they are governed by other organizations besides OSHA or because they are relatively low risk based on their intended use. This exemption means that containers of these substances do not require an OSHA-required label (which does not mean they do not need a label at all), do not need to be added to the Haz Com Chemical Inventory, and do not need an SDS on file. The two primary exemptions are office supplies that are used as intended and household products. Office supplies (e.g., ink within pens) do not need to be included in the Haz Com Program. Household products covered by the Consumer Product Safety Commission that are used IN A HOUSEHOLD MANNER AND KEPT IN HOUSEHOLD QUANTITIES do not need to be included in the HazCom Program. Examples are provided below for the household exception.

- Two containers of Windex window cleaner and a bottle of multi-surface cleaner, stored under the breakroom sink, used sporadically
  - Exempt, becausethese are household products used and stored in a household manner
- Bulk quantity of containers of window cleaner used daily by custodians
  - Not exempt, because it is stored in excess of typical household and it is used daily for long periods of time unlike it is in typical households
- A pallet of windshield wiper fluid containers used only when needed
  - Not exempt, because stored in excess quantities as compared to a typical household
- A single bottle of nail polish remover which is used in its entirety to remove paint from metal
  - Not exempt, because this is not intended household use

Employee Information and Training

Employers shall provide employees with effective information and training on hazardous chemicals in their work area at the time of their initial assignment and whenever a new chemical hazard the employees have not previously been trained about is introduced into their work area. Information and training may be designed to cover categories of hazards (e.g., flammability and carcinogenicity) or specific chemicals. However, chemical-specific information must always be available through labels and SDSs.

At a minimum, employee training shall include the following:

- The location and availability of the written HazCom Program, including the required list of hazardous chemicals and SDSs;
- Methods and observations that may be used to detect the presence or release of a hazardous chemical in the work area (such as monitoring conducted by the employer, continuous monitoring devices, visual appearance or odor of hazardous chemicals when being released, etc.);
- The physical, health, simple asphyxiation, combustible dust, and pyrophoric gas hazards, as well as hazards not otherwise classified, of the chemicals in the work area
  - Physical hazard—a chemical that is classified as posing one of the following hazardous effects: explosive, flammable (gases, aerosols, liquids, or solids), oxidizer (liquid, solid, or gas), self-reactive, pyrophoric (liquid or solid), self-heating, organic peroxide, corrosive to metal, gas under pressure, or in contact with water emits flammable gas.
  - Health hazard—a chemical that is classified as posing one of the following hazardous effects: acute toxicity (any route of exposure), skin corrosion or irritation, serious eye damage or eye irritation, respiratory or skin sensitization, germ cell mutagenicity, carcinogenicity, reproductive toxicity, specific target organ toxicity (single or repeated exposure), or aspiration hazard.
- The measures employees can take to protect themselves from these hazards, including specific procedures the employer has implemented to protect employees from exposure to hazardous chemicals, such as appropriate work practices, emergency procedures, and personal protective equipment to be used; and,
- The details of the HazCom Program developed by the employer, including an explanation of the labels received on shipped containers and the workplace labeling system used by their employer; the SDS, including the order of information and how employees can obtain and use the appropriate hazard information.

REVIEW QUESTIONS

1. What is the three-worded phrase commonly used to refer to Hazard Communication and how does it relate to HazCom?
2. Are employees who travel to different work locations throughout their shift required to have a Hazard Communication Program physically with them at all times?
3. When is a label NOT required on a chemical container?
4. What does an exclamation point with a red diamond mean (as a pictogram)?

5. If a chemical does not come with an SDS, does that mean you don't need one? What needs to happen?
6. At what times are employees required to be trained on HazCom?
7. What is the difference between a physical hazard and a health hazard?

REFERENCE

Occupational Safety & Health Administration [OSHA]. (2013). Regulations (Standards-29 CFR 1910.1200). Retrieved from https://www.osha.gov/laws-regs/regulations/standardnumber/1910/1910.1200

Postface

A Hilarious Bonus Story

Based on a true story (really).... A safety inspector shows up at a coal mining jobsite. The inspector gets out of his truck, and a few employees reluctantly greet him. One employee, Jim, is wearing neither his required hard hat nor steel-toed boots. After an introduction, the inspector points at Jim and asks, "Where's your hard hat and steel toes?" Jim offers some profane language and walks away. The next day the inspector shows back up to the jobsite to continue the inspection. The inspector parks next to a job trailer, and the supervisor greets him as he gets out of his truck. During the exchange, the trailer door flies open with a loud bang. Jim walks out of the trailer, WEARING NOTHING BUT A HARD HAT AND BOOTS, COMPLETELY NAKED, and says, "I got your hard hat and safety shoes right here buddy!" The inspector doesn't say a word, gets back in his truck, and drives away.

Index

About the Author

Elliot Laratonda has worked in various industries as a safety and health (S&H) professional, including construction, manufacturing, warehousing and distribution, academia, consulting, and research. Elliot is a certified safety professional. Elliot received his BS in Applied Safety Health and Environmental Sciences from Indiana University of Pennsylvania. He obtained his MS in Industrial Systems Engineering (Human Factors and Safety Engineering focus) from Virginia Tech University. As an S&H professional, Elliot's focus has been OSHA compliance and the development and implementation of S&H programs. Elliot has implemented S&H programs and initiatives that have contributed to a significant reduction in incident rates and a measurable improvement in safety culture. Elliot has been an active member and committee chair for the American Society of Safety Professionals, and he has taught industrial safety to undergraduate engineering students. At the time of this book's publication, Elliot is working as a safety adviser and consultant.

CONTRIBUTING AUTHORS

Paige Laratonda is a certified safety professional and a graduate of Indiana University of Pennsylvania (IUP) with a BS in safety, health, and environmental applied sciences. Paige obtained her certified safety professional credential five years after graduating from the IUP EHS Applied Sciences program and has been working in the safety field ever since. Paige is an active member of ASSP and currently works full time in higher education. She is a safety coordinator at Penn State University where she consults on OSHA and safety program compliance throughout the campus. Paige also assists where necessary to assess risk and provide recommendations for hazard controls in research. Paige has other safety, health, and environmental experience in industries such as manufacturing, heavy equipment, mechanical contracting, and nonprofit companies.

Travis Spagnolo is a certified safety professional (CSP) who graduated from Indiana University of Pennsylvania with a degree in Safety, Health, and Environmental Applied Sciences. Travis has also earned a master's in Business Administration from Ohio Dominican University. Travis currently works as a project safety manager for Turner Construction Company in their Columbus, Ohio, office. In this role, Travis is responsible for managing the safety and health of Turner employees and all of Turner's trade partners and subcontractors on a large-scale construction project in central Ohio. Prior to joining Turner in December 2021, Travis spent the first seven years of his career working for Safex, Inc., a health and safety consulting company in Westerville, Ohio. At Safex, Travis supported a variety of clients in construction, manufacturing, and general industry settings. Travis is experienced in performing site and facility audits and gap assessments, developing lockout/ tagout procedures, conducting fall protection assessments, and evaluating confined spaces and developing entry procedures. He is also an experienced safety and health trainer. Travis is experienced in developing and delivering training for topics that include fall protection, lockout/tagout, confined space entry, scaffolding, silica, and various other safety and health topics. Travis serves on the construction safety planning committee for the Bureau of Workers' Compensation Ohio Safety Congress and Expo. For six years, Travis served as a judge for the Builder's Exchange of Central Ohio's Safety Achievement Awards. Travis is a member of the American Society of Safety Professionals and holds OSHA 30-Hour Construction and OSHA 40-Hour HAZWOPER cards. He is also an OSHA ten- and thirty-hour general industry authorized trainer. In 2021, Travis was one of the thirty-eight safety professionals from around the world to be named one of the National Safety Council's "Rising Stars of Safety."

Ingram Content Group UK Ltd.
Milton Keynes UK
UKHW041534300323
419412UK00024B/240